Organic Farm Management Handbook 2023

Edited by:
Nic Lampkin, Mark Measures and Susanne Padel

Research:
Rowan Dumper-Pollard and Kathrin Heimbach

Design, production and contributor:
Phil Sumption, Bio Communications

12th Edition (October 2023)

Published by:
Organic Research Centre
Trent Lodge, Stroud Road, Cirencester, Gloucestershire, GL7 6JN
Tel: +44(0)1488 658298
Email: hello@organicresearchcentre.com
Website: www.organicresearchcentre.com

ISBN: 978-1-3999-6622-1

Individual copies: £35 + post and packing

Preface to the twelfth edition (2023)

Much has happened since the last edition of the Organic Farm Management Handbook. The UK organic market has seen consistent growth, which was strong during the Covid-19 lockdown, but subsequently levelling off, partly as a result of higher living costs in the UK. The illicit invasion of Ukraine by Russian armed forces has shocked us all and has disrupted the organc and conventional grain markets in the UK and in Europe, with a period of high domestic prices that since then has stabilised again.

Organic land area increased again during the last 5 years in England, but growth remains slower than growth in retail sales. This indicates opportunities for the future. Many organic farms (small growers, family farms and estates, selling into different market outlets) have continued to maintain successful organic businesses and most farm types have remained as profitable as conventional. Dairy has experienced disruption because of competition from high conventional milk prices and high organic cereal prices have also affected other livestock producers.

Despite the emerging popularity of alternative farming approaches in conventional farming, such as regenerative, organic farming continues to be the gold standard for a more sustainable, agroecological way of farming, uniquely offering legal standards in the UK with the retained EU regulation, and good consumer awareness and well-developed markets. Organic farming offers solutions to addressing the increasingly urgent problems of climate change, biodiversity loss and deteriorating human health facing the world. With clearly evidenced environmental and social benefits, the UK is well placed to benefit from an expansion of the organic sector. Changes in the support schemes, such as the new Sustainable Farming Initiative in England, indicate opportunities. However, one of the key barriers to further uptake is the lack of current information on the business performance of organic farms and related management issues that this handbook aims to fill. Good decisions need to be based on more than back of the envelope calculations that only consider headline prices. The sensitivity analyses provided for each enterprise allow for the selection of prices and yields that may be applicable to the particular circumstances of an individual farm.

Our thanks are due to all the researchers, contributors, and sponsors who have helped with the production of this handbook.

Rowan Dumper-Pollard, Nic Lampkin, Mark Measures and Susanne Padel, 2023

Preface to the first edition (1994)

The idea for an Organic Farm Management Handbook has been several years in the making, but 1994 seems an opportune moment for it to be realised. The growth in organic farming throughout Europe and the introduction of financial support for it in the United Kingdom and Ireland in 1994, mean that organic farming is on the brink of a major new phase in its development. Sound management information will be a key factor in this, supported by the increasing availability of first-hand data from Defra-funded and other research programmes.

We have attempted to compile data from a range of sources to provide best possible estimates for the physical and financial performance of organic crop and livestock enterprises. Our initial estimates have been considered by a panel of experienced organic producers and modified in the light of their comments. It is inevitable, however, that in a publication of this type, particularly the first issue, some errors will creep in. As editors, we must accept full responsibility for these, but we would be grateful for feedback if calculation errors are noted. We would also welcome feedback on the content and style of the different sections.

We would like to acknowledge the willingness to assist and helpful comments received from organic producer members of British Organic Farmers/Organic Growers Association, as well as other producers and researchers who commented on the initial drafts of this handbook. We owe a big debt to the researcher/contributors Tony Pike, Hugh Bulson, and Christopher Stopes who did much of the ferreting around for information on the earlier versions of the handbook. Thanks are also due to Susanne Padel for advice and comments on the technical content. Finally, we acknowledge financial support from the EC for the collection of financial data under research contract AIR 3-CT92-0776.

Nic Lampkin & Mark Measures, July 1994

Please note: on the 2nd of October 2023, shortly after the 2023 Organic Farm Management Handbook was sent to print, a new payment rate was announced for organic top fruit.

This now means that all revenue claim payments for "OR5: Organic conversion - top fruit" will be £1,920 per hectare for up to three years (full information available at: www.gov.uk/countryside-stewardship-grants/organic-conversion-top-fruit-or5). This affects information provided on page 51, where payment rate for "OR5: Organic conversion - top fruit" is £1,254 per hectare which was the figure provided at the time of print. The increase in payment is welcome as the production costs for organic top fruit farms have risen significantly in recent times. Please keep this new payment rate in mind when using the handbook.

Acknowledgements

We gratefully acknowledge the generous financial support of Organic Farmers & Growers (OF&G) for this edition (see main advert on back cover). We have included advertising from a number of additional sponsors throughout the handbook. Their support has been critical to publishing this edition of the handbook. We also thank the many individuals and companies who supported our campaign and helped make the Handbook happen.

We would also like to acknowledge the support of all the individual farmers, companies and government agencies who have provided us with information to make this edition of the Organic Farm Management Handbook possible, in particular the following individuals and organisations:

- Graham Redman and James Webster (Andersons) for permission to use data from the Nix Farm Management Pocketbook 2023 and support with the fixed costs.
- Rebecca Laughton, Andrew Burgess (Produce World), John Pawsey (Shimpling Park Farm), Andrew Trump and Dominic Amos (Organic Arable), William Waterfield (Farm Consultancy Group), Tom Wood (Robin Appel), Nigel Mapstone (Mole Valley), Mike Thompson (For Farmers), David Coleman (Terrington Farm Machinery), Daniel Bridger (Premier Plants), Peter Plate (Endell Veterinary Group), Iain Tolhurst (Tolhurst Organic Partnership), John Twyford (The Dairy Group), Joy Greenall, Duncan Hendry (Cope Seeds), Paul Totterdell (Cotswold Seeds), Harriet Bell (Riverford), Sally Westaway, Libby Bird (For Farmers), Paul Muto (Oliver Seeds), Michael Marriage (Doves Farm), Nick Walton (Bagthorpe Farms), John Newman (Abbey Home Farm), Ben Pearce (Shillingford), Jordan Lilley and John Overvoorde (Delfland Nurseries), Roger Kerr and Steve Clarkson (OF&G), Christopher Stopes (EcoS Consultancy Ltd & OF&G), Helen Browning, Adrian Steele and Sarah Hathaway (Soil Association), Ali Capper (Stocks Farm), Richard Plowright (Plowright Organic), Andrew Skea (Skea Organics), Helen Wade (Eastleach Downs Organic Farm), Clive Hill (OLMC), Sean Ruffel (Organic North).
- Philippa Hall, Dr. Lindsay Whistance, Dr. Will Simonson, Janie Caldbeck, Matt Smee, Dr. Julia Cooper and other staff from the Organic Research Centre.

Front cover photo: Kay Ransom

Contents

Why an Organic Farm Management Handbook?

Organic agriculture offers solutions to addressing many of the current problems agriculture is facing. 'Organic' is one of the most recognized food labels globally, and most people in developed countries including in the UK consume some amount of organic food. There is legal recognition of the standards, the market for organic food continues to grow, globally and in the UK, even if high inflation has recently led to some stagnation. But producers in the UK are not taking up the opportunities that this growing global market offers.

There is much interest in alternatives to intensive agriculture, such as agroecology and regenerative. While regenerative agricultural systems may 'restore' or 'regenerate' natural ecological functions, an accepted definition is lacking[1]. Many practices now advocated as regenerative have their origin in organic farming, practiced by over 3,000 producers in the UK. The term 'regenerative' itself is attributed to an organic pioneer. Robert Rodale in the US described it as an holistic approach to farming that encourages continuous innovation and improvement of environmental, social, and economic measures. Organic and regenerative agriculture have some things in common, in particular the goal of improving soil health. This was expressed by one of the pioneers of the organic movement in Britain, Lady Eve Balfour in her 1943 book *The Living Soil*. The statement "*The Health of Soil, Plant, Animal and Man is one and indivisible*" is now an important part of the Principle of Health, one of the four principles of organic farming of IFOAM Organic International. Organic farming is the best developed of various alternative approaches, with a good market infrastructure and growing producer base globally, internationally agreed legal standards and a track record of successful farm businesses delivering multiple outputs including high quality food, environmental protection and high biodiversity.

So why are not more farmers going organic? One of the key barriers is the lack of current information on the costs and business performance of organic farms and related management issues that is vital to anyone contemplating the seismic shift to organic farming. The purpose of this handbook is to address this gap by providing relevant and up-to-date business management information for farmers, advisers and others concerned with the management of organic farms.

---

1. https://www.tabledebates.org/building-blocks/what-is-regenerative-agriculture

The information presented in this handbook is based on historical data and estimates for the coming season, representing what an average farmer, given adequate resources, might expect to achieve. The data have been derived from different research reports and estimates of industry experts and have been reviewed by experienced organic advisers and producers. It is important to note the gross margin data throughout do not represent the performance of representative surveys of organic farms. Such data does not exist for the UK as a whole. Some data for a number of farm types in England have been included in Section 5.

One of the most striking observations when reviewing the performance of organic businesses is the very wide variation between farms and between years. In using the data for budgeting purposes, it is necessary to take the circumstances of the individual farm into account. Food and agricultural input and output prices and currency exchange rates are changing regularly, adding a considerable level of uncertainty to any forward projections. EU-exit and the war in the Ukraine have massively added to this uncertainty. As in previous editions, all gross margin data include a sensitivity analysis indicating how margins might be affected by changes in input and output prices and by different levels of productivity. Whilst we have attempted to make best possible estimates, the editors cannot be held responsible for their application in specific circumstances.

This handbook is designed for use in conjunction with similar handbooks for agriculture in general. For this reason, data not relating specifically to organic farming have only been included where particularly relevant, e.g., information on agricultural and environmental support programmes. Labour and machinery costs for individual operations are only covered briefly, as reliable data on these aspects of organic systems remain scarce. For further information on these topics, reference will need to be made to other publications.

What is organic farming?

Organic farming aims to be sustainable by adopting agroecological[2] approaches to system management, making more effective use of the farm's own resources in preference to external inputs. Organic standards and regulations help realise these ideas in practice, with guidance on specific permitted or restricted practices and inputs, and certification to enable access to premium markets.

---

2. Lampkin N et al. (2015) The Role of Agroecology in Sustainable Intensification. Report for Scottish Natural Heritage. Organic Research Centre, Newbury. https://orgprints.org/id/eprint/33067/

Organic farming can be defined as an approach to agriculture where the aim is:

to create integrated, humane, environmentally and economically sustainable agricultural production systems. Maximum reliance is placed on locally or farm-derived, renewable resources and the management of self-regulating ecological and biological processes and interactions in order to provide acceptable levels of crop, livestock and human nutrition, protection from pests and diseases and an appropriate return to the human and other resources employed.

In many European countries, organic agriculture is known as ecological or biological agriculture, reflecting this reliance on ecosystem management and living organisms rather than external inputs[3].

The objectives of health, quality and sustainability lie at the heart of organic farming and are among the major factors to be considered in determining the acceptability of specific production practices. The term 'sustainable' is used in its widest sense, to encompass not just conservation of non-renewable resources (soil, energy, minerals) but also issues of environmental, economic and social sustainability. The term 'organic' is best thought of as referring not to the type of inputs used, but to the concept of the farm as an organism, in which all the component parts – the soil minerals, organic matter, micro-organisms, insects, plants, animals and humans – interact to create a coherent and stable whole.

The key characteristics of organic farming practice include:

- protecting the long-term fertility of soils by maintaining organic matter levels, encouraging soil biological activity and careful mechanical intervention;
- providing crop nutrients indirectly, using relatively insoluble nutrient sources which are made available to plants by the action of soil micro-organisms;
- nitrogen self-sufficiency through the use of legumes and biological nitrogen fixation, as well as effective recycling of organic materials including crop residues and livestock manures;
- weed, disease and pest control relying primarily on crop rotations, natural predators, diversity, organic manuring, resistant varieties and limited (preferably minimal) thermal, biological and chemical intervention;

3. Lampkin and Padel (1994) The Economics of Organic Farming–An international perspective. CAB International, Wallingford; P.4. See also IFOAM's definition: www.ifoam.bio/why-organic/organic-landmarks/definition-organic

- the extensive management of livestock, paying full regard to their evolutionary adaptations, behavioural needs and animal welfare issues with respect to nutrition, housing, health, breeding and rearing;
- careful attention to the impact of the farming system on the wider environment and the conservation of wildlife and natural habitats.

These ideas have been captured in the four core principles of organic agriculture defined by the International Federation of Organic Agriculture Movements (IFOAM)[4]:

Health: Organic agriculture should sustain and enhance the health of soil, plant, animal, human and planet as one and indivisible;

Ecology: Organic agriculture should be based on living ecological systems and cycles, work with them, emulate them and help sustain them;

Fairness: Organic agriculture should build on relationships that ensure fairness with regard to the common environment and life opportunities;

Care: Organic agriculture should be managed in a precautionary and responsible manner to protect the health and well-being of current and future generations and the environment.

Organic farming in the United Kingdom and in Europe

Certified organic production

In the UK, the total area of organic and in-conversion land was 508,585 hectares in December 2022 with small increases in land area since 2018. This represents 3.0% of the total agricultural area (excluding common grazing) in the UK.

After a period of decline, organic land area very slowly increased again in the last 5 years in England, but not in the other three home nations. Growth in organic land area in the UK remains slower than growth in retail sales and is considerably slower than in Europe. In addition to 3,285 producers at the end of 2022 there were 223 producer/processors and 1,988 organic processors certified.

In Europe, the number of certified holdings increased from 6,700 in 1985 to approximately 340,000 in 2014. In addition to producers there are 51,000 processors and 1,900 importers. The certified land area in Europe increased from less than 100,000 hectares in 1985 to 11.6 million hectares in 2014 . In the EU, 5.7% of land area was managed organically; 40% of organic land area is permanent pasture and 35% of land was used for arable crops.

4. www.ifoam.bio/why-organic/shaping-agriculture/four-principles-organic
5. See www.organic-europe.net

Number of organic farms and total land area (ha) registered by Defra (end 2022)

Country	No. of Producers	Organic area (ha)	In-conversion area (ha)	Total organic (ha)	% total utilised agric area
England	2 202	292 725	19 636	312 361	3.4%
Scotland	379	92 500	18 400	110 900	2.3%
Wales	552	75 505	2 122	77 627	4.7%
Northern Ireland	152	7 529	168	7 697	0.8%
Total Dec 2022	**3 285**	**468 259**	**40 326**	**508 585**	**3.0%**
Total Dec 2021	3 401	464 674	41 974	506 648	2.9%
Total Dec 2020	3 407	457 640	31 318	488 958	2.8%
Total Dec 2019	3 494	457 115	28 064	485 179	2.7%
Total Dec 2015	3 429	500 763	20 635	521 398	3.0%
Total Dec 2010	4 741	667 551	50 794	718 345	4.2%
Total Dec 2005	4 343	544 124	87 020	631 144	3.6%
Total Dec 2000	3 563	194 171	187 416	527 323	

Source: Defra Statistics annually from 2009 (published middle of following year); previously UKROFS.

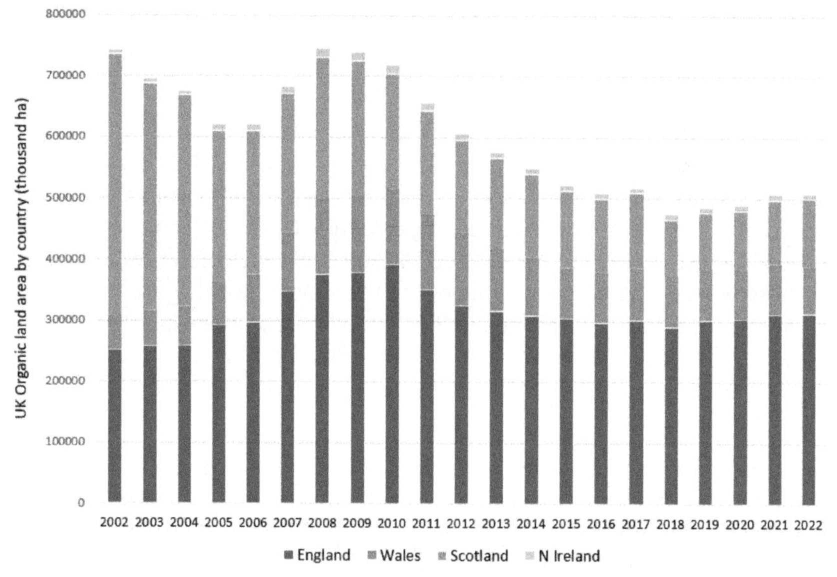

UK organic and in-conversion land area by nation (kha), 2002-2022

Source: Defra (2022) and previous years[6]

6. www.gov.uk/government/collections/organic-farming

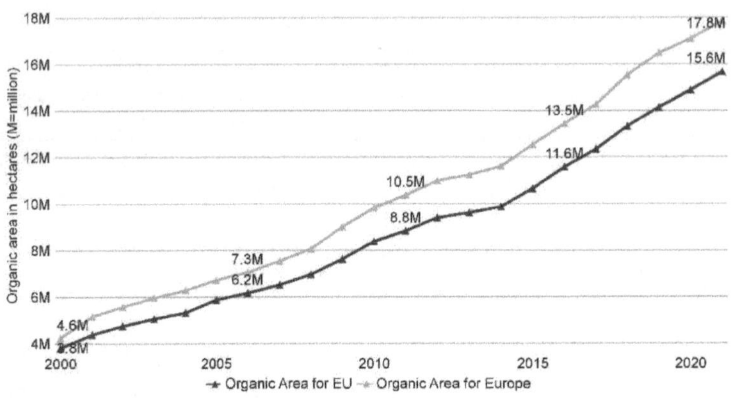

Certified organically-managed and in-conversion land area in Europe (EU, EFTA, CEE), 2000-2022 (year end)

Source: FIBL at www.organic-europe.net

The nature of the organic farm business

The concept of organic farming as a whole farm system, not just a set of 'organic' practices or individual 'organic' enterprises, is fundamental to understanding organic farming's performance as a business.

For example, the issue of yields under organic management often leads to concerns that organic farming cannot be viable financially. This is not always based on actual evidence of yields on organic farms but on misconceptions and incorrect assumptions, such as "organic farming is a return to pre-1950s farming." Organic yields are taken to be the same as yields under zero nitrogen input use yields, where no other aspect of the system is changed (see Figure below). Also, claims have been made that three times as much land area would be needed to produce organic crops compared with conventional, on the mistaken assumption that nothing is produced during the fertility building phase of organic rotations.

In assessing the performance of organic farming, there is a need to consider the restructuring of the farm system and not just modifications to agricultural practices, such as the level of input use or the combination of individual inputs. This is a strategic rather than tactical change involving a large number of different variables. The organic system relies on both management (redesign of the system, for example of the crop rotation) and alternative inputs (for example labour, machinery, organic manures etc.) to replace those that have

been withdrawn. In the context of the traditional production response curve (below), changes in factors other than the target variable input will lead to a shift to a different production function. The prohibition of synthetic nitrogen fertilisers in organic farming (and certain other inputs) and the emphasis instead on biological nitrogen fixation, represents just such a shift to a different production response curve, not a shift along an existing conventional one to the zero-nitrogen input level.

This is well illustrated by data on yield responses to nitrogen use in grass-only swards compared with clover/grass mixtures. Not only are the yields under zero N use substantially higher in the grass/clover case, but it may actually be more profitable, depending on input costs, to choose this zero N option than to fertilise a grass-only sward. If nitrogen contained in manures and slurry applied to the land (typically around 100 kg N/ha) is also taken into account, the potential to reduce costs and maintain output under organic management can be significant.

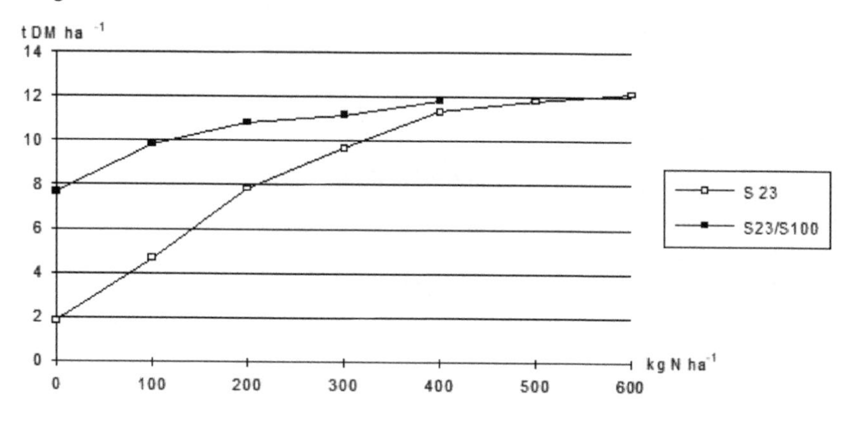

Nitrogen fertiliser production response curves for perennial ryegrass only (S23) and perennial ryegrass/white clover mixtures (S23/S100)

Source: Thomas, C., Reeve, A. and Fisher, G. E. J. (eds.) (1991) Milk from grass. 2nd edition. British Grassland Society, Reading

The relatively low total productivity differences for farms with forage-based livestock mean that there may be potential to maintain profitability through cost reductions alone, especially in the context of rising fertiliser and energy prices. Most crop and intensive livestock enterprises such as free-range pigs and poultry will require higher prices (see Section 2) in addition to lower costs, in order to compensate for reduced yields and the higher costs of some organic inputs, such as concentrate feeds and seeds.

The achievement of similar gross margins simply on the basis of lower costs would require a percentage reduction in costs much greater than the percentage reduction in yield. For example, if in the high-intensity conventional case, variable costs account for 40% of output and output falls by 40%, then costs would have to fall to zero to compensate, which is clearly not feasible. If variable costs are only reduced by 50%, as is typical for organic cereal crops, a 33% premium would be needed to maintain the same gross margin. Premiums of more than 50%, currently achievable for many crops, will result in higher financial output and higher gross margins than on conventional farms. On the other hand, where conventional production is already less intensive, as in the case of cereal production in the Less Favoured Areas, and output falls by only 10%, variable cost reductions may be sufficient on their own to compensate for the lower output.

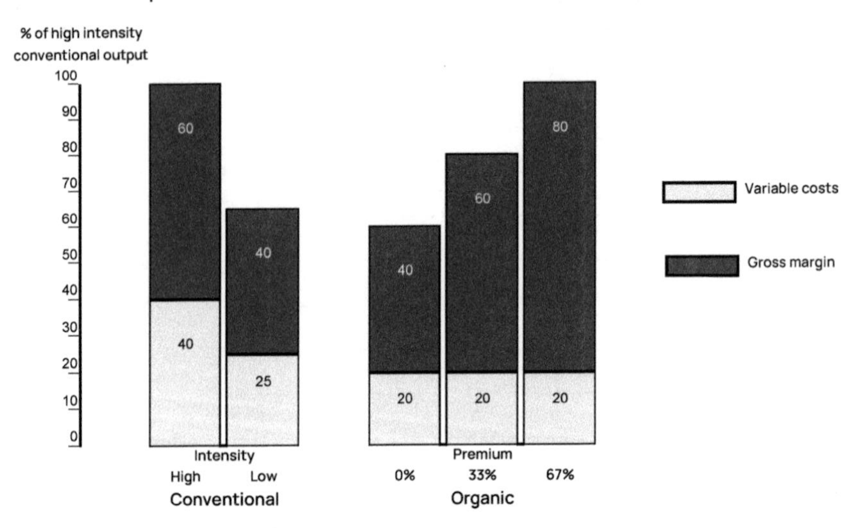

Relationship between output, variable costs, premium prices, and gross margins in organic and conventional agriculture

The yield differential experienced in the UK between conventional and organic arable and vegetable crops is highly variable, dependant on a number of factors including the level of input use, crop type and disease pressure. Typically, average organic field-scale root crop, vegetable and cereal yields produce 50 -60% of standard, high input conventional yields, although the differential is less for oats and rye. The limiting factor for organic yields is generally in the first instance nitrogen supply; ways of making use of nitrogen fixation by legumes and improving recycling and nitrogen utilisation are priorities for all farms.

A different picture emerges if not only yield of individual crops and enterprises but total productivity is considered. An analysis of the total productivity of organic farms compared with similar conventional farms carried out as part of the Defra-funded monitoring of organic farm incomes in England and Wales, indicated that for livestock and mixed farm types, total productivity differences were only about 10%, although for specialist cropping holdings the differences can be much greater.

Organic yields (t/ha) from Farm Business Survey data[7] for England 2019-2022

Product	Unit	Organic average (3 years)	2021/22	2020/21	2019/20
				Sample farms	
Winter wheat	t/ha	4.0	3.5 (11)	3.8 (10)	4.6(13)
Spring wheat	t/ha	3.6	3.8 (7)	3.9 (5)	3.2 (6)
Spring barley	t/ha	3.1	2.5 (24)	3.2 (31)	3.6 (35)
Winter oats	t/ha	4.4	4.5 (16)	4.3 (14)	4.5 (19)
Spring oats	t/ha	3.5	3.3 (17)	3.8 (19)	3.5 (19)
Field beans	t/ha	2.3	2.3 (9)	2.3 (11)	2.4 (9)
Milk	litres/cow	6170	5981 (35)	6249 (39)	6249 (40)
Stocking	cows/ha	1.3	1.3	1.3	1.3
Milk	litres/ha	7774	7656	7811	7849

Several recent scientific publications reviewing the evidence regarding organic yields worldwide concluded that organic crop yields are on average 75-80% of conventional yields, but that there is significant regional and crop type variation, with organic yields ranging from 20% to 177% of conventional and only 5% differences for rainfed legumes and perennials. For developing countries organic yields can be up to 30% higher and therefore relevant to food security. This is likely to be a result of lower use of inputs in such countries. What is clear from all these studies is that yield differences found for specific crops in specific regions cannot be generalised globally[8].

7. Organic Farming in England at https://www.ruralbusinessresearch.co.uk/
View ORC reports on organic farm incomes: http://tinyurl.com/OFIreports
Mind the gap – exploring the yield gaps between conventional and organic arable and potato crops https://orgprints.org/id/eprint/31232

8. For references and further detail see Lampkin, N. et al. (2015) The Role of Agroecology in Sustainable Intensification. Report for Scottish Natural Heritage. Organic Research Centre, Newbury. https://orgprints.org/id/eprint/33067/

Gross margins and organic farming systems

Gross margins, such as those presented in this handbook, are useful for farm planning, for example prior to a conversion to organic farming. They are also used for comparisons of enterprises, for example between different enterprises on the same farm, between different organic holdings or between conventional and organic enterprises. However, there are some important limitations to their use:

- Gross margins represent the difference between enterprise output and variable costs, excluding whole farm or fixed costs. Gross margin comparisons between enterprises with different fixed cost structures can be very misleading, particularly where conventional variable costs (fertiliser and crop protection inputs) have been substituted by fixed costs (machinery and labour) in the organic context;
- Certain inputs such as organic manures, lime and phosphate are applied on a rotational basis with residual effects on subsequent crops, so it is unrealistic to expect their costs to be carried by the individual enterprise to which they are first applied;
- High individual enterprise gross margins for cash crops do not reflect the potentially very different enterprise mix on organic farms and in particular the need for fertility-building crops in the rotation. Cereals and other cash crops tend to account for a much lower proportion of land area than on conventional farms, while temporary grass, fodder crops, grain legumes and potatoes and vegetables are relatively more important. Taken out of the whole farm context, individual gross margins can therefore be very misleading.

The interactions between enterprises in an organic farming system make an important contribution to the overall financial performance of the system, particularly where external inputs are no longer available to fulfil a particular function. Grazing livestock on organic farms provide a means of utilising the fertility-building phase of a rotation that is usually based on forage legumes. The forage enterprises can therefore be seen as part of the 'costs' of achieving high returns for some crops like wheat and potatoes. In a similar way, although beef cattle may give lower returns per hectare than sheep, the parasite control and grazing management benefits from interactions between the two species are such that the beef enterprise can, in certain situations, be seen as a justifiable 'cost' of sheep production.

It is therefore often inappropriate to consider the economics of a single enterprise, such as organic wheat, unless the context of the whole farm system

is considered. The optimisation of individual elements within a system does not necessarily result in the optimisation of the whole system: the interactions between enterprises lie at the heart of the holistic, systems approach to agriculture to which organic farming aspires. However, a balance has to be achieved between the economic benefits of increased diversity and integration of different enterprises (exploitation of complementary relationships, more even distribution of labour requirements, increased stability of the system, reduction of risk) and the benefits obtained from economies of size and specialisation particularly on smaller holdings.

Further reading

Lampkin, N (2020) *Organic Farming*. Chapter 11 in Soffe R and Lobbly M (ed.) The Agricultural Notebook. 21st edition. Wiley-Blackwell, Oxford.

Organic farming: Basic principles and good practices. FIBL Dossier 2021. www.fibl.org/en/shop-en/1141-organic-farming-principles

Further reading sections are included throughout this handbook. They include specialist technical guides available from several organisations. In addition, codes of practice in the standards documents of the various certification bodies operating in the UK and Ireland provide guidance (see Section 3). For information on other general organic farming publications see also Section 14.

Market developments and organic consumers[1]

The organic market has grown consistently for the last ten years. The most recent annual figures available show a 1.6% increase in sales of organic food and drink, making the UK organic market worth £3.1 billion in 2022. Growth had been particularly strong during the Covid-19 pandemic but has slowed down over the last two years with high inflation and the cost-of-living crisis. The retail sales value was up 25.4% for 2022 compared to before the pandemic (2019). This illustrates the importance of taking more of a long-term view. The values for 2023 will be published in March 2024.

Trends in the main global markets

At a European level, the most recent figures for many countries (from 2021[2]) show the organic market to be worth nearly €54.5 billion (EU €46.7 billion). In 2022 Germany remained the country with the largest market (€15.3 billion retail sales), followed by France (€12 billion) and Italy (€4.6 billion). Between 2012 and 2021, the value of the EU's organic market more than doubled. Both France and Germany have well-established national logos for organic products and substantial government investment in generic promotion of organic food. The EU now has a target of 25% of land area being organic by 2030 that is also expected to encourage policies supporting market development in the Member States.

Trends in the European Market for 2022 overall are expected to show stagnation or even decline, making 2023 and beyond difficult to predict. 2022 was characterised by inflation following the war in Ukraine as well as being the first year with 'almost no Covid-19' conditions, when out-of-home consumption returned to pre-Covid-19 levels and grocery purchases for use at home reduced. The German organic market overall showed a small decline (-3.5%) in 2022, with the greatest reductions in sales observed in natural food stores and other outlets (e. g. butchers and bakers). Sales in supermarkets (including discounters) increased in the same year by 3.2%. The German retail sales value in 2022 was higher than in the pre-Covid-19 year of 2019. There has

1. This section was prepared using information from the Soil Association's 2023 *Organic Market Report* together with other published and forthcoming reports, and consultation with key industry stakeholders. 2023 retail sales figures will be published in the 2024 Soil Association Organic Market Report in March 2024.
   Statistics for organic production are published annually by Defra.
   www.gov.uk/government/statistics/organic-farming-statistics-2022/
2. www.fibl.org/fileadmin/documents/shop/1254-organic-world-2023.pdf

been a significant increase in purchase volume among Austrian consumers, with a rise of 31.2% from 2019 to 2022, and in terms of value, there has been a notable increase of 42.7%. One key factor contributing to the consistent sales of organic food in Austria are the relatively lower price increases compared to conventional food. In 2021, organic food prices only rose by 7.5%, in contrast to 11.5% price increase for conventional food.

In 2021, the UK was the 8th largest exporter globally of organic products to the EU, falling back to 12th position in 2022.

After the EU, the next largest market for organic products, including UK exports, is North America, where organic retail sales reached €53.9 billion in 2021. The USA reported retail sales of €48.6 billion and Canada €5.3 billion in this period. Data from the US Organic Trade Board indicates that the USA market grew by 4% in 2022, despite the impacts of inflation and the ending of the pandemic stimulus to growth.

Many other countries, in particular in the Middle East and Asia, are also proving attractive for UK exporters. Further data on global organic markets are available in the annual World of Organic Agriculture report, published by FIBL and IFOAM[3].

Trends in different UK market outlets

Trends differ between different market outlets. *Multiple retailers* remain the most important sales outlet for organic food in the UK, accounting for 61.8% of total organic sales. Towards the end of 2022, organic sales in multiple retail returned to growth (+0.5%). *Home delivery* makes up 18% of organic sales and *independent retail* 13.8%. Strong growth was reported for *food service/catering* which accounted for 6.3% of organic sales in 2022. Unlike other European countries, the UK has fewer *specialist organic shops* and dedicated supermarket chains, which have been important outlets elsewhere. Multiple retailers have been important for growth of the organic market in several countries. In Germany for instance, supermarkets have become the driving force in the organic market, whereas specialised retailers have faced competition.

Many organic product categories in the UK have experienced growth between 2018 and 2022. Dairy was up 34.7% dairy and fruit by 11.2% from 2018 to 2022. Between 2021 and 2022 there was some decline in sales for fruit, veg and ambient grocery, meat and some other categories, but sales of dairy and babyfood and drink increased. Negative growth from 2021 to 2022 illustrates the impact of the cost-of-living crisis and inflation on the organic market.

3. https://statistics.fibl.org/index.html

Other organic products have become more popular, such as hot beverages and crackers, with a 5-year growth of 33.3% and 18.8%, respectively. Herbal tea and rice cakes have been especially popular.

To support sustainable growth within the organic sector, it is important to maintain a strong UK organic production base throughout the supply chain, including processing. According to Defra, the number of processors has decreased over the last ten years. Organic producers have needed to work as closely as possible with retailers and processors to communicate the organic message, as well as embracing online marketing and social media and investing more in direct sales.

Organic consumers

Organic customers have very different levels of commitment to buying organic food and also different levels of engagement. This ranges from regular organic shoppers who are committed to the values of organic food and a broader range of ethical issues (including local food, environment, animal welfare and fair trade), to those that occasionally buy one or two organic products. Many consumers only purchase organic products in certain key categories, with fruit & vegetables, meat, milk, and yoghurt remaining the most popular. Market growth can come from increasing the spending on organic food of regular customers as well as encouraging occasional shoppers to spend a bit more.

The top reasons for buying organic food include 'less use of pesticides', 'health benefits', 'better taste', 'environmental benefits', 'animal welfare benefits', and 'better quality' and these motivations have changed little over time. Consumers with a high sustainability awareness have been found to be older, wealthier and members of smaller households. Tight budgets, as experienced during the cost-of-living crisis, may impact the willingness of consumers to pay more for organic food and on the volume per buyer. People may spend less, visit supermarkets less often and reduce their purchasing frequency. But some of the changes recently observed may also reflect post-Covid-19 returns to more out of home consumption and less shopping for preparing meals at home and less home-baking.

Organic consumers[4] are likely to be passionate about their food as well as having a strong ethical and environmental conscience. However, the high price of organic produce remains an often-quoted reason by consumers for not choosing organic. On average, consumers pay a premium of 19.3% for an

4. See Padel (2016) Introduction to global markets and marketing of organic food. In Deciphering Organic Foods: A Comprehensive Guide to Organic Food Consumption" (Karaklas I, Muehling D (Eds). Nova Publishing, Hauppauge. http://orgprints.org/30798/

organic product compared to the conventional equivalent. Although for many consumers a higher price is justified, the 'high price' image remains a significant obstacle for market development that needs to be addressed by carefully targeted marketing and information sharing. Likewise, the limited availability of organic products remains a barrier to growth. As most organic sales take place in supermarkets, if key organic products are not available in these stores, consumers are likely to buy the non-organic alternatives, including premium lines, Fairtrade or locally sourced rather than looking for alternative sales outlets that stock organic items. Finally, for consumers, purchasing organic produce is not a black-and-white issue. Organic is not an isolated concept in their mind but is connected to other concerns and values about food.

Covid-19 had an undeniable impact on the organic market, both positive and negative[5]. Consumers recognised the importance of their own health and personal well-being and were willing to spend more money on potential health benefits. Consumers have been found to perceive organic goods to be more 'natural' and thus healthier. The multiple lockdowns and pandemic measures changed the shopping habits of UK consumers and some of these changes continued in the post-Covid-19 era. For example, the use of online purchasing. Over one in four organic products are now purchased online.

Market outlets

Multiple retailers

Sales through multiple retailers account for around 64% of total organic sales and are worth around £1.9 billion. They declined by 2.7% in 2022 after a growth of 2.4% in 2021. Waitrose's Duchy Originals brand has helped to boost organic sales and has celebrated its 30th birthday as the UK's largest own-label organic food and drink brand. It furthermore has launched a collaboration with Planet Organic to increase their organic product range in 2023 in both stores and online. Morrison's recently relaunched its organic range to communicate their commitment to sustainability more clearly to their customer base.

The rapid growth of the discounters Aldi and Lidl has also created a new market for organic produce with a limited organic range offered by both supermarkets. This pitches organic produce at price-conscious consumers and has shown success in the last years. Across all sales channels, only discounters managed

5. Wang H, Ma B, Cudjoe D, Bai R & Farrukh M. (2021). How does perceived severity of COVID-19 influence purchase intention of organic food? British Food Journal, 124(11), 3353–3367. https://doi.org/10.1108/BFJ-06-2021-0701

to grow their organic share in 2022. Their parent companies have experienced success selling organic products in other European countries and the US and will be aiming to repeat this in the UK. Aldi and Lidl have made strong British sourcing commitments, so may look to source much of this organic produce from British farmers but no official commitment has been published yet.

Producers supplying any supermarket need to recognise the professional requirements that are associated with securing a stable relationship. Quality, reliability, and commitment are necessary to retain supermarket buyers in this competitive and often very difficult market. Producers should also be prepared for further rationalisation of the supply base as supermarkets aim to consolidate the number of product lines they stock to improve efficiency.

Other sales channels

Independent retailers: accounted for £453.1 million organic retail sales in 2022, which is a decline of 3.3% compared to 2021. This market comprises of mainly single outlet businesses as well as some organic retail chains with a limited number of stores. Box schemes and mail order/home delivery as a category of retail outlet, includes both independent organic outlets (such as box schemes) as well as online services of major retailer chains. Online sales channels grew strongly during Covid-19 and have remained an important outlet since, making up 18.9% of total sales and are worth £558.6 million. Organic has a higher presence in online sales, compared to conventional products and customers pay the highest organic premium on average online compared to other sales channels. Over a 5-year period (from 2018 to 2022), online sales have managed to grow from 12.4% to 18.9% of organic retail sales. Online sales have been essential for the growth of the total organic market in the last five years. Most consumers (63%) still shop in different places, part online and part in stores.

Abel & Cole and Riverford remain the two most dominant independent online retailers for home delivery. Smaller box schemes continue to struggle in many instances, often due to their geographic isolation and not being able to match the product range of their larger competitors. Some box schemes are also seen as not flexible enough for some customers. Establishing a very close relationship between producer and consumer and focusing on local supply can be a key strength of the smaller producer-owned schemes. New enterprises that offer solutions for software and logistics open-up possibilities for an easier online market entry for farmers and simplify the management and administration significantly.

Farmers' markets: There are over 500 farmers' markets held in the UK, with a wide variation in the number of stalls and customers. The share of Farmers' markets for total organic sales is small and precise data are not available, but for many producers, farmers' markets are an important place for promotion in their local community.

Farm shops: A number of organic farms have opened farm shops. They experienced similar hardship as other independent retailers and their sales have dropped by 3.3% in 2022. Like all direct sales, they require regular commitment, time and a willingness to engage with people. But they can also offer great rewards through meeting consumers more directly. Inflation is seen as one of the biggest challenges for smaller independent shops, especially combined with the cost-of-living crisis.

Community Supported Agriculture (CSA)[6] allows communities to participate in the development of food initiatives in a partnership between farmers and consumers, where responsibility and rewards are shared. The basic practice is for members of the community to 'buy-in' to the local farm in return for a supply of farm products over the year, usually on a not-for-profit basis. Different models exist and numbers are increasing, partially due to initiatives such as the Soil Association's CSA Network. This indicates a shift in the relationship that people want with their food and those who grow it. The key difference to box schemes is that people become members or shareholders and, in some cases, pay an annual subscription to the farm and may be involved in decision-making for what is grown.

Catering and public procurement

Catering: Organic sales for out-of-house consumption in 2022 were estimated at £195.5 million in the UK, a year-on-year increase of 152%. This rapid increase is attributable to Pret-a-Manger's boom in business after lifted Covid-19-restrictions and their use of organic ingredients in their hot drinks. Even though catering establishments fall outside the scope of the regulations, they are expected to be able to provide evidence to Trading Standards showing that they are sourcing organic ingredients. Voluntary certification for catering establishments is offered by the Organic Food Federation (OFF), Organic Farmers and Growers (OF&G) and Soil Association Certification Limited (SA Certification) – see Section 3 for certifier details.

6. www.soilassociation.org/csa for the online CSA Action Manual and other information. See also: https://communitysupportedagriculture.org.uk/resource-a-z/

Like most other sales channels, supplying to catering outlets requires a continuous investment of time, energy and capital to meet the specifications required by caterers.

Public procurement[7]: The idea that organic operators develop direct supply lines to schools, hospitals and local authorities fits well with organic principles, but entry into this market means overcoming existing barriers in procurement procedures, contract arrangements, service expectations (e.g. due to kitchen equipment) and competing against other locally-produced food. Some successful pilot schemes exist, and the number of outlets that serve some organic food has increased through the Food for Life Catering Mark. Over half of English primary schools now have a Soil Association Catering Mark, with the gold and silver catering mark requiring a small proportion of organic produce to be on the menu. However, spending cuts lead to tighter purchasing budgets which may limit further development. There are opportunities for farms to get involved in working with the local school community, offering farm visits and information for staff and pupils to learn and experience how their food is produced.

Processing

The supply of UK products to UK processors remains low with the supply chain being dominated by imported products. Producers wanting to supply this market must be professional in managing production and delivery to suit the needs of the market. They must recognise that processing cannot be seen as a dumping ground for substandard or out-of-spec produce. Processors will only set aside their facilities for organic food if they see an opportunity to process a reasonable quantity, to justify the cost and effort of separation and cleaning down.

---

7. www.organic.aber.ac.uk/schools; www.soilassociation.org/foodforlife

Market developments during 2022/23[8]

This section, mainly aimed at producers, summarises key trends in market development for important product categories and current prices. Forward planning of marketing is important, particularly at difficult times. However, market price trends are always difficult to predict as they are influenced by seasonal trends as well as retailer and consumer demand. Inflation and the development of the war in Ukraine has had a significant impact on production costs and prices. The outcome of this crisis is unclear, as of now, as is the political support for organic farming post-EU-exit, making predictions about organic trends uncertain.

Meat and eggs

Beef: The supply is influenced by organic feed costs along with the conventional price for store and finished cattle, as this influences organic farmers' willingness to sell into non-organic outlets. Generally, organic meat, fish and poultry has experienced a stronger decline with a negative growth of −9.7% in 2022. However, in the summer of 2023, the market is resilient and organic beef prices do not experience the same price declines as in the conventional sector and are able to maintain their premiums.

In 2022, a considerable number of organic beef farmers reverted to non-organic farming primarily due to high feed costs. Consequently, the availability of organic beef cattle, particularly barren cows, became increasingly scarce. The decline in numbers has reached a point where certain abattoirs have been unable to slaughter any organic cattle due to insufficient quantities to come up with a viable kill plan. The availability of beef remains limited, and abattoirs are unable to meet their weekly orders, leading to a higher number of organic dairy beef going back to non-organic. Prices are at £5.40/kg DW in the summer of 2023 and have had up to 90p premium in the previous months.

Lamb: Demand for lamb in the UK is generally declining, as consumers increasingly favour poultry to red meat and tighter consumer budgets impact consumption choices. Prices in the summer of 2023 were £6.30/kg DW with a premium of 30-40p/kg. Overall, domestic price trends are likely to reflect both the domestic balance of supply and demand for organic lamb and export potential in the conventional market. Organic prices are expected to closely follow conventional, with a 0-5% premium.

---

8. Based on 2023 *Organic Market Report*, Soil Association and other sources. For regular updates see www.soilassociation.org/farmers-growers/market-information/

Pork and bacon: Organic pig production is effectively divided into two sectors: larger enterprises supplying the multiple retail sector and smaller enterprises producing (and sometimes processing) for local markets and direct sales. The organic sector reflects trends in the wider British pork sector. The conventional British breeding herd reached an all-time low in 2022 and general pig numbers have decreased by 2.5%. Similarly, the small UK organic pork sector has become even smaller. The Ukraine crisis highlighted the dependence of this sector upon imported organic feed. Many organic pig farmers drastically reduced their numbers or dropped out completely for financial reasons. Sales further dropped in 2022 with inflation as they did for beef and poultry. There were small increases in the organic herd size for pigs in 2021 and 2022 according to Defra, but numbers for 2023 are expected to show a decrease. The sector's reliance upon imported organic pig feed means that the cost of production is closely linked to the value of the Pound Sterling and is therefore likely to increase.

At a European level, consumer demand has slowed whilst production has increased, leading to more of a balance between supply and demand compared to previous years. Importantly, China remains a lucrative market for those who can supply in enough volume.

Table poultry: The number of organic chickens had increased in 2021 but declined in 2022. Multiple severe waves of avian flu inhibited any further growth and weakened organic poultry systems through trade restrictions, supply disruptions, increased costs and culling. Consumers increasingly prefer chicken over red meats due to perceived health advantages and affordability, but the market has not managed to recover fully in 2022.

Eggs: Organic egg sales decreased slightly in 2022, following three years of strong growth from 2019-2021. At the end of 2022, sales had started to pick up again by 21.2%, raising hopes of a full recovery. The egg market in the UK experienced a crisis in November 2022 with retailers allegedly refusing to pay more for eggs following the rising costs of farmers due to the Ukraine crisis and Avian flu. The latter amplified shortages, with organic and free-range farmers being forced to keep their stock inside and not being able to sell their eggs for their usual premium.

Milk and dairy produce

Organic dairy products continue to be the largest product category in the organic market, accounting for 26.9% of sales in the multiples in 2022. Dairy was one of only two organic product categories to increase sales in 2022 (the other one being baby food and drink). And it is one of the four sectors that has grown most consistently over the last five years (alongside fruit & veg, chilled convenience and alcohol).

The UK organic milk market includes milk buying groups (Organic Herd, previously OMSCO is the largest group) and direct contracts with dairy companies (e.g., Arla). Despite growth of the organic dairy market in 2022, the sector faced some challenges. Feed and energy costs shot up due to the Ukraine war increasing the cost of production and as supply chain issues stemming from the Covid-19 crisis continued. This resulted in some producers going out of dairying and others reverting to conventional, leading to a drop in farm gate milk production since 2022. Meanwhile, consumers were less willing to spend a premium on liquid milk due to the rising popularity of milk alternatives and tight budgets. With conventional milk facing similar struggles, the farm gate premium for organic milk decreased from 10p/l to 6p/l but conventional prices have fallen significantly during the last year due to high costs and undersupply. Liquid milk and yogurt saw some volume losses in multiple retail in 2023, which have affected organic sales slightly more than conventional[9].

Arla has increased their presence in the organic milk market and aims for 50% growth in their organic dairy business in the next five years. It has introduced its own standards over and above the national organic standard that came into effect in January 2022 (see Section 3). All organic producers working with Arla have to take steps to enhance biodiversity and transition to 100% renewable energy. They are also expected to reduce their carbon dioxide equivalent (CO_2e) emissions per kilogram of milk by 30% by 2028, ahead of the company's objectives that aim for a similar reduction by 2030.

Organic Herd (previously OMSCO) has intensified their focus on the US export market and made antibiotic-free milk obligatory for their members to meet the tougher USDA organic standards.

Farm gate prices differ between different buyers. In the summer of 2023, the organic premium accounted for between 5 and 10 p/l. We have used an average farm gate price of 43p/l in this edition, but it all depends on the buyer and it should be noted that some buyers have a relatively flat pricing structure whilst others vary their prices according to the season and may range from 36p/l to 45p/l. In the summer of 2023 Organic Herd paid 49p/l, down from a peak of 51p, Muller paid 49p, down from 56p and Arla 5p premium at 40p, down from 55p. There is also variation in how quality is rewarded.

Key to sustainable growth and relative price stability in the organic sector is balancing input costs and keeping a high premium for organic milk.

9. https://ahdb.org.uk/news/organic-deliveries-decline-as-consumers-continue-trading-down

At the same time, buyers are calling for better communication of the quality differences between conventional and organic to consumers. The market is increasingly influenced by conventional advertising promises that are similar to organic regulations, which undermines the advertising message of organic. Therefore, the whole industry (producers, Control Bodies, milk buyers and milk processors) will need to agree a future strategy for how to achieve the stability and resilience they had pre-Covid-19 again in the long-term.

Export markets are an important outlet for UK organic dairy products, including the Middle East, South Korea and particularly the USA.

Vegetables and fruit

Vegetables and fruit (fresh produce) is the second largest category within organic foods and drink in value terms, accounting for 20.7% of organic sales in multiples in 2022. Alongside milk, they are also generally the first products that new organic consumers start to buy. Most of fresh fruit and vegetables (including salads) is sold unbranded.

The pandemic led to a surge in demand for veg box schemes as consumers sought alternatives to traditional shopping methods. The concerns around food safety, reduced access to supermarkets, and a desire to support local producers and food systems prompted many individuals and families to change their consumption habits during 2020 to early 2022. With the rising cost-of-living and inflation in late 2022 and 2023, some consumers have switched towards cheaper options.

Over the past five years, the total organic and in-conversion horticultural area in the UK (vegetables, fruit, nuts and potatoes) has increased slightly to 12,600 hectares. Of those 1,500 hectares are currently under conversion. Given a small decline in the area over the past ten years, it is not expected that the UK area for fruit and vegetables will increase significantly in the near future.

Cereals and other grains

Markets have been improving as a result of lack of supply of certain key raw materials arising from a number of factors. Whilst demand continues to grow for organic cereals for human consumption and animal feed, the domestic supply does not keep up, resulting in heavy reliance on imports – over 50% for grains during 2018-2022. Over these last five years, organic cereals have seen a strong increase in cultivation, growing from 37,400 hectares in 2018 to 49,500 hectares in 2022 according to the latest data from Defra.

The 2022 harvest season was heavily impacted by the Ukrainian war and its substantial effects on grain and fuel markets. Record prices characterized this period, with milling wheat surpassing £500 per tonne. Many arable growers benefited, especially since the pre-inflation prices of seed and diesel hadn't affected crops planted in autumn 2021. Generally, crop yields were high, and the majority of grains didn't require drying after harvest.

However, the high prices led to a drop in demand in 2022, particularly as livestock farmers contended with rising feed costs. Many dairy farmers opted to grow their own grain or turned to forage, reducing milk yields, resulting in an overall decline in demand of around 20%. The barley market faced its challenges too, with unsold barley even at significant discounts to wheat.

For the 2023 harvest, farm gate prices are expected to be around £300 for feed wheat, while barley prices are expected to be around £25 lower and feed oats a further £10 lower.

As we go to press, a range of factors are impacting prices at once and the political and economic climate is very uncertain.

Accessing the organic markets

However, market access cannot be taken for granted. Both retailers and processors compare price, quality and continuity of supply between the different suppliers both in the UK and elsewhere. The future prospects of the organic market should not be judged by recent headlines alone, but should take into consideration more long-term trends. Producers also need to consider the relative profitability of organic farming and not only market developments.

Working collaboratively can help to achieve better access to markets through pooling of raw materials and economies of scale. Very few businesses in the agrifood sector are large enough to be able to have the required volume and consistency of supply and sufficient bargaining power on their own, and the same is true in the organic sector. In the UK, there are number of organic producer groups that engage in marketing, such as Organic Arable, Organic Herd (previously OMSCO), Calon Wen, Meadow Quality and Organic Livestock Marketing Cooperative that work for producers to gain better access to organic markets.

It remains essential to discuss intentions with any future buyer, be it producer group, processor or retail buyer, to secure outlets prior to starting production. For organic meat, milk, cereals, fruit or vegetables, access to market outlets is far from automatic and all organic producers need to continuously maintain and develop their own markets. Planning for the financial viability of organic production should also include considering the impact of price fluctuations, particularly when specific markets are over-supplied. All gross margins shown in the handbook therefore include a sensitivity analysis (see Section 7) that allows the impact of price variation at the margin to be looked at. In addition, the following issues should be considered:

- careful planning of investment;
- getting to know the end-consumers through market research among real and potential customers;
- access to up-to-date market intelligence information;
- developing good relationships with customers – whether supermarket buyers, wholesalers or consumers – and establishing supply agreements or contracts where possible;
- planned production to anticipate and respond to trends in market demand;
- product innovation;
- co-operation with other producers to reduce costs, share expertise, and strengthen the producers' position in the market.

Organic food promotion

Every organic farm/business should engage in marketing and promotion of its products and most farms have fantastic stories to tell[10]. Some organisations offer support and marketing toolkits.

Ooooby (www.ooooby.com), amongst other providers, offer software support for producers who want to set up their own box scheme and can assist with marketing, logistics, sales analysis and all tools needed for a functioning online shop.

The Organic Trade Board (OTB)[11] is a membership organisation for organic businesses that has the mission to grow organic sales in the UK and does this through work in communicating the benefits of organic to consumers via long term integrated promotion campaigns, advising and helping businesses and working with them to grow the market, working with all retail channels to promote organic products and influencing government to do more for organic food and farming.

---

10. Frost et al (2015) Communcating organic food values. OCW https://orgprints.org/id/eprint/51585/
11. www.organictradeboard.co.uk/

Promotional events for organic food

BioFach: 13-16 February 2024 (and similar dates annually), Nürnberg Exhibition Centre, Germany. One of the biggest events with over 35,000 visitors, 2,765 exhibitors and an accompanying congress programme. Now touring with BioFach Global in Tokyo, Shanghai, Delhi , Riyadh and Sao Paulo. Further details: Nürnberg Messe GmbH Messezentrum; D-90471 Nürnberg, Germany. Tel +49 (0) 9 11 86 06-0 www.biofach.de/en

Natural & Organic Products Europe: 14-15 April 2024, Excel Centre, London. The UK's largest annual trade event for the natural products and organics industry. Details from Diversified Business Communications UK (DBC UK), Third Floor, Blenheim House, 119-120 Church Street, Brighton BN1 1UD. Tel: 01273 645 126 or 01273 645 120. See www.naturalproducts.co.uk

Wake up to Organic: This OTB campaign aims to show consumers how easy it is to switch to organic products at breakfast and provides an opportunity for independent stores, cafes and restaurants to run individual promotional campaigns, coordinated around a specific day. After a break in the last few years due to Covid-19 restrictions the event returned in June 2023 with over 50 shops and brands participating across the country.

Organic September: A month of co-ordinated activities and promotion allowing producers and retailers to stimulate interest in organic food and farming. A wide range of events take place across the UK and many businesses participate with specialist organic offers. For further details contact the Soil Association.

BOOM Awards (The Soil Association's Best of Organic Market): The award aims to showcase organic food in the UK. It is open for entry from producers, retailers and suppliers in different categories (e.g. taste, special product, producer of the year, best local food initiative, best box scheme). For 2024 applications contact the Soil Association.

There are also many regional and local food events.

Rural development grants for marketing and processing

Policy support for farmers and for rural development continues to be set by the four devolved administrations. They have held separate consultations regarding future development priorities and thus changes in the new regulations differ between the administrations (see also Section 5). Before the EU-exit, the majority of policy support in this area was provided through the EU Common Agricultural Policy (CAP).

In **England**, a first round for applications for a competitive grant on *Adding value - Processing and Marketing* as part of the *Farming Investment Fund* was open in 2022. Successful applicants have been notified to submit a full application for grants from £25,000 to £300,000 by 31 Jan 2024. No information on future rounds is available at the time of printing[11]. The following two schemes may also be relevant in future. Some initiatives in local communities could be considered as part of the UK *Shared Prosperity Fund*, which is administered by Local Authorities. In Areas of Outstanding natural Beauty (AONBs) the *Farming in Protected Landscape Scheme* may include support for nature friendly and sustainable farming in those regions along with other outcomes, administered through the local protected landscape bodies.

In **Scotland**, grants are managed through the Rural Payments and Services system and are likely to remain aligned to schemes offered under the CAP but following national priorities, such as Good Food Nation Vision. The Food Processing, Marketing and Co-operation Grant Scheme (FPMC) aims to support Scotland's journey towards becoming a Good Food Nation. The programme operators in a single year modus, all projects must be completed in the same financial year as the grant was awarded. It covers various support capital and non-capital costs. Email: FoodProcessingGrant_Enquiries@gov.scot

In **Wales**, the three-year programme of transition to the Sustainable Farming Scheme includes the Agricultural Diversification Scheme, a capital grant scheme available to farmers, aiming to support the establishment of new viable agricultural diversification enterprises on farms, the development of existing novel or niche enterprises, and developing diversified income streams. Applications closed in January 2023, future rounds possible but no details have been announced. Further information from Wales Rural Network Support Unit can be viewed here: https://businesswales.gov.wales/walesruralnetwork/

In **Northern Ireland**, schemes operating under the CAP are now closed and no new grants related to food or farm diversification have been announced.

Sources of further information on Rural Development Programmes

England: www.gov.uk/rural-development-programme-for-england

Scotland: www.gov.scot/farming-and-rural/

Wales: www.gov.wales/farming-countryside

Northern Ireland:
www.daera-ni.gov.uk/grants-and-funding/rural-development-grants

---

12. https://www.gov.uk/guidance/farming-investment-fund#adding-value-grant

Further reading

Market information and lists of certified processors and retailers are available from Control Bodies (for contact details see Section 3).

Organic Market Report of the Soil Association: Bristol. Download at www.soilassociation.org/marketreport (free to SA licensees). The 2023 report was published in March.

KANTAR x OTB Annual Organic Market Report 2023, Organic Trade Board. (Free to OTB members). www.organictradeboard.co.uk/members-hub/login/

Introduction to global markets and marketing of organic food. By Susanne Padel In: Deciphering Organic Foods: A Comprehensive Guide to Organic Food Consumption" (Karaklas I, Muehling D (Eds). Nova Publishing, Hauppauge. http://orgprints.org/30798/

What you can say when selling organic food. Summarising approved advertising claims. Soil Association, Bristol www.soilassociation.org/certification/food-drink/business-support/marketing-organic/

Farmer Consumer Partnerships – How to successfully communicate the values of organic food. Handbook for producers. Free download at http://orgprints.org/17852

Websites

Soil Association Marketplace: www.soilassociation.org/farmers-growers/market-information/organic-marketplace/

Organic Farmers and Growers: https://ofgorganic.org/

Organic Trade Board:www.organictradeboard.co.uk/

Organic Arable Marketing Group: www.organicarable.co.uk

National Farmers' Retail and Markets Association (FARMA): www.farma.org.uk

Business intelligence for the organic food industry: www.ecoviaint.com/

Defra Organic Farming Statistics www.gov.uk/government/collections/organic-farming

Organic farming is recognised and regulated globally. In 2022, 74 countries had fully implemented regulations on organic agriculture.

Organic farming production standards in the UK

All food products sold as organic must by law follow certain standards (both in the UK and in the EU). All operators have to be regularly inspected and certified by approved certification (control) bodies. For agricultural holdings certification can also be a condition of eligibility for any the organic farming support schemes throughout the UK and the Republic of Ireland (see Sections 4 and 5).

After the UK exit from the EU the organic provisions in the UK are covered in Part 5 of the Agricultural Bill of 2020.

The **Organic regulations in force in the UK** are the retained EU Council Regulation 834/2007[1] and the retained Commission regulations 889/2008 and 1235/2008. These Regulations are no longer in force in the EU and have been replaced by a new organic Regulation (EC/848/2018 with several amendments and implementing rules) in force since January 2022 (see below). The scope of these Regulations covers all operators along the whole organic food supply chain; only catering is exempt. Private standard owners can have stricter rules.

Defra is overseeing the legal basis for organic standards for the whole of the UK, in consultation with the organics Four Nations Working Group (FNWG). An Expert Group on Organic Production (EGOP)[2] has been established that provides comprehensive advice on future organic policy and technical issues requiring specific expert advice. Defra is also assisted by the UK Accreditation Service (UKAS) in licensing the UK organic certification (control) bodies and for overseeing their inspection activities. Northern Ireland will continue to follow EU regulations.

The retained Council Regulation (EC) 834/2007 on organic food sets out objectives and principles of organic production as well as basic requirements for plant and livestock production, aquaculture, compound feed, the preparation of products (i.e., processing), and criteria for the approval of substances. The retained Commission Regulation (EC) 889/2008 lays down more detail on the implementation for crop and livestock production, incl. aquaculture, seaweed and yeast production, processed products, wine, packaging, transport and storage, labelling, and inspection, and Annexes listing permitted inputs.

---

1. All EU legislation can be found at eur-lex.europa.eu: Search by document number and year (e.g., Year: 2007; Number: 834 or Year: 2008; Number: 889)
2. https://www.gov.uk/government/groups/expert-group-on-organic-production-egop

The statutory conversion period starts at the earliest when the operator has notified their activity to a control body. During the conversion period all the rules have to be followed, but the products cannot be marketed as organic. The statutory conversion period for arable and grassland is at least two years before sowing of the crop to be harvested as organic, or, in the case of grasslands at least two years before its use as organic feed or at least three years for perennial crops, such as orchards.

The control system requires an annual verification of each operator, carried out by organic control/certification bodies . This then leads to certification. All UK certification bodies have to be accredited by UKAS, the UK Accreditation Service in accordance with the requirements for bodies operating product certification systems (ISO 65 or EN 45011). Irish Certification Bodies operating in the UK are accredited in Ireland, but also carry a UK number although this is under review.

Some certification bodies in the UK operate their own standards, which may include higher specifications regarding livestock husbandry, environmental requirements, and the use of certain inputs. Producers are advised to consult with their inspection bodies about specific requirements and can obtain copies of the full standards from them. Regular updates to the standards are distributed to licensees in the form of newsletters. Some sector bodies have or are aiming to develop standards in new areas currently not regulated in detail, such as catering, pullet rearing and transplant production. Co-operation agreements between some certification bodies exist.

Organic farming regulations in the EU[5]

The first organic regulation in Europe came into force in 1991 (EEC No 2092/91) which was amended several times and fundamentally revised in 2007. After a further revision a new **European Regulations (2018/848)** for organic food came into force on 1 Jan 2022, replacing the second common European Regulation (EC/834/2007 and implementing rules). UK Operators wishing to export to the EU have to comply with this new EU regulation, which also applies in Northern Ireland. The new European regulation aims to strengthen the European organic market and increase consumer trust in the organic brand. Changes include some clarification and broadening of the scope, but mass catering remains outside. New requirements address for example a further reduction of non-organic feed and increasing the percentage of feed from the farm for livestock, an increase in

---

3. See also www.gov.uk/guidance/organic-food-uk-approved-control-bodies
4. See for list of accredited producer certification bodies: www.ukas.com/find-an-organisation/browse-by-category/?type=279
5. For details see ec.europa.eu/agriculture/organic/eu-policy/legislation_en

the use of organic seed, also covering heterogenous materials (e.g., populations). The rules strengthen the need for production in the soil and place more emphasis on reducing the risk of contamination with non-organic products. Accreditation remains mandatory for certification bodies. New rules have also been included for the import of organic produce from non-EU/EEA countries, which now also apply to imports from the UK. Importers to the EU will have to prove that their products meet EU organic standards and ensure that they have been certified by a recognized certification body. Furthermore, the use of the EU label is mandatory for all organic products sold within the EU, independent of origin.

In the **Republic of Ireland**, the Organic Farming Unit established by the Department of Agriculture, Food and the Marine is responsible for all aspects of organic production including the implementation of EU regulations. Producers can either register directly with the Department or with any of the three registered certification bodies: Irish Organic Farmers and Growers Ltd. (IOFGA); Organic Trust Ltd, and Demeter Standards (Ireland) Ltd.

Objectives, principles and key rules according to UK and EU Regulations

The main objectives of organic farming[6] can be summarised as follows:

- establish a sustainable management system for agriculture methods;
- aim at producing products of high quality;
- and aim at producing a wide variety of foods and other agricultural products that respond to consumers' demand for goods produced by the use of processes that do not harm the environment, human health, plant health or animal health and welfare.
- The new EU regulation also lists protection of the environment and the climate, maintaining fertility of the soils, contribution of high levels of biodiversity, encouraging short distribution channels and local production, and fostering the development of organic plant breeding as objectives. Some objectives have been rephrased as principles.

Organic farming principles[7] can be summarised as follows:

- the appropriate design and management of biological processes based on ecological systems using natural resources, living organisms and mechanical methods;
- the restriction of the use of external inputs;
- the prohibition of the use of chemically synthesised inputs or strict limitation to exceptional cases.

6. Article 3 of Regulation EC/834/2007; Article 4 of Regulation EC/848/2018

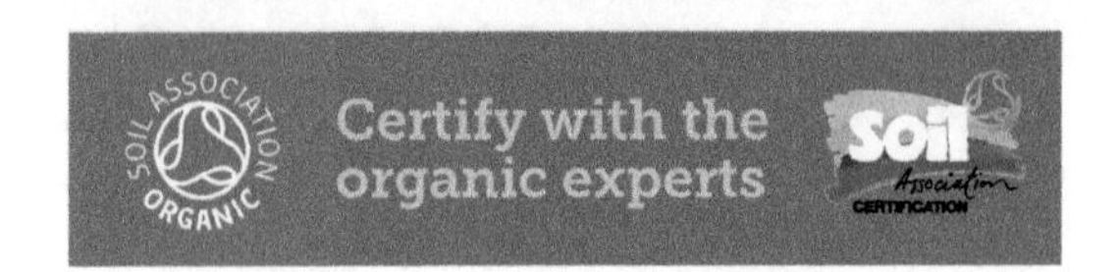

The principles allow that the rules can be adapted to certain specified or local conditions, but very few exceptions have been granted under this rule. In addition, specific principles are defined for farming, processing of organic food and processing of organic feed.

The use of GMO in organic products is prohibited. To determine whether an input is GM free, the operator can rely on the labels in line with the Directive 2001/18/EC. This principle will also continue to apply, even if the use of New Genetic Techniques will be liberalised.

The use of permitted inputs is regulated in the relevant sections for crops, livestock and processing with lists of permitted inputs in the Annexes of Regulation (EC) 889/2009. Agricultural raw materials should mainly be of organic origin, with some exceptions that are set out in the relevant sections of this Handbook. It is strongly recommended that producers consult their certification body, if they are uncertain about the conditions under which a certain input can be used, as some UK certification bodies require consultation prior to use.

International Federation of Organic Agriculture Movements

Contact address

Charles-de-Gaulle-Str. 5, 53113 Bonn, Germany

Tel: +49 (0) 228 92650 10

Contact: Denise Godinho, Membership & Communications Manager

Email: communications@ifoam.bio

Website: www.ifoam.org

IFOAM–Organics International has been leading, uniting and assisting the Organic Movement since 1972. As the only global organic umbrella organization, IFOAM advocates organic agriculture as a viable solution for many of the world's pressing problems. With 800 affiliates in over 120 countries, the organisation campaigns for the greater uptake of organic agriculture by promoting its effectiveness in preserving biodiversity and fighting climate change. IFOAM also offer training courses, provide services to standard owners, certifiers, operators, and implement organic programmes.

Standards:

The IFOAM Norms for Organic Production and Processing, including the IFOAM Standard for Organic Production and Processing, the IFOAM Accreditation Requirements for Bodies Certifying Organic Production and Processing, and the Common Objectives and Requirements of Organic Standards (COROS), serving as a basis for the approval of standards into the family of standards. The standards are revised regularly, the last free download publication dating from August 2014.

7. Article 4 EC/834/2007; Article 5 of 848/2018

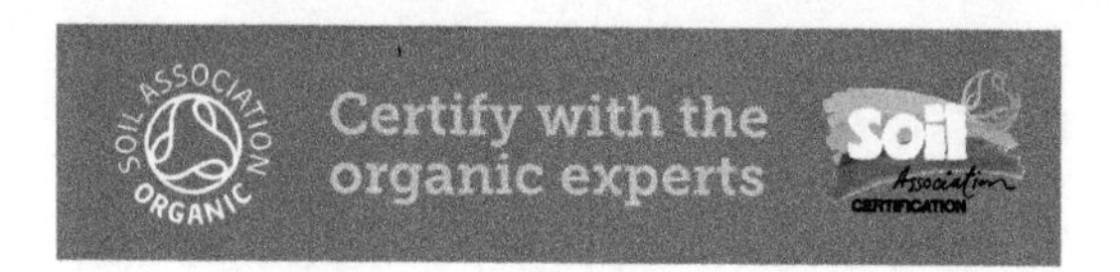

Accreditation service:
Certification bodies can be accredited through the IOAS. For further information contact from International Organic Accreditation Service (IOAS). Neither IFOAM nor IOAS directly certifies individual farms or businesses.

Contact address:

IOAS, 119 2nd Ave. West, Dickinson, ND 58601, USA
Tel: +1 701 483 5504 Fax: +1 701 483 5508
Email: info@ioas.org
Website: www.ioas.org

Organic Farmers & Growers C.I.C.

GB-ORG-02

Contact address

Old Estate Yard, Shrewsbury Road, Albrighton, Shrewsbury SY4 3AG.

Tel: 01939 291800

Email: info@ofgorganic.org

Website: www.ofgorganic.org
For certification in scotland visit: www.sopa.org.uk/ for more information. Certification by OF&G Scotland.

Standards: Paper copy £12.00 +VAT (free pdf download)

Inspection and certification charges: Charges may be revised at any time, so please contact the office for up-to-date information.

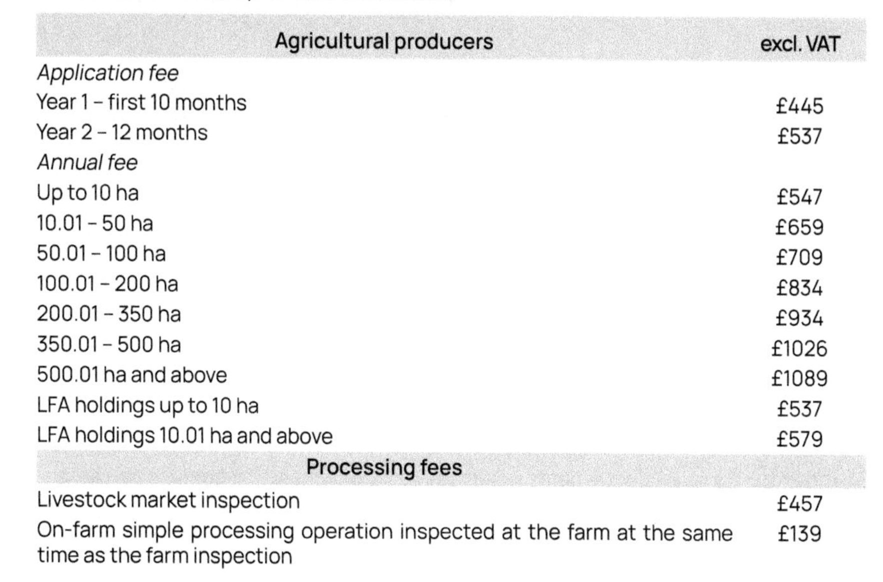

Agricultural producers	excl. VAT
Application fee	
Year 1 – first 10 months	£445
Year 2 – 12 months	£537
Annual fee	
Up to 10 ha	£547
10.01 – 50 ha	£659
50.01 – 100 ha	£709
100.01 – 200 ha	£834
200.01 – 350 ha	£934
350.01 – 500 ha	£1026
500.01 ha and above	£1089
LFA holdings up to 10 ha	£537
LFA holdings 10.01 ha and above	£579
Processing fees	
Livestock market inspection	£457
On-farm simple processing operation inspected at the farm at the same time as the farm inspection	£139

For all other processing fees see Form RD15 on website.

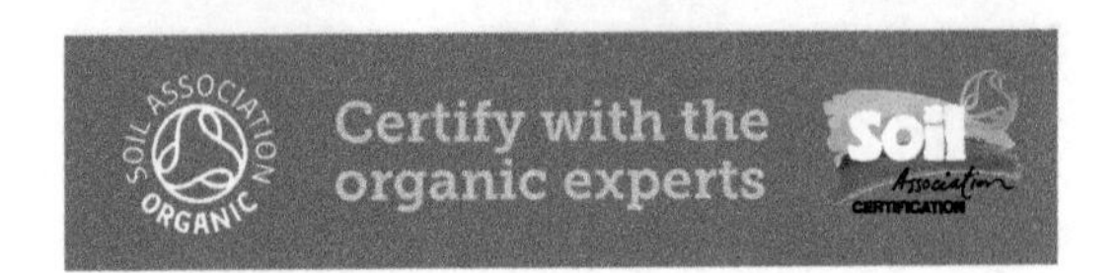

Working in partnership with Acoura, OF&G is able to provide farm assurance and organic inspections that can normally be carried out on the same day: Red Tractor Assurance for Farms, Dairy Scheme, Beef & Lamb Scheme, Crops & Sugar Beet Scheme, Pigs Scheme and Fresh Produce Scheme. Farm Assured Welsh Livestock (FAWL in Wales).

Organic Food Federation

GB-ORG-04

Contact address

31 Turbine Way, Eco Tech Business Park, Swaffham, Norfolk
PE37 7XD

Tel: 01760 720 444

Email: info@orgfoodfed.com

Website: www.orgfoodfed.com

Standards: Compliant with current EU legislation for processing, production, aquaculture, importing, trading and warehousing.

Inspection and certification charges:

Membership subscriptions vary according to the status and type of operation. Each application is judged on its merits. Charges are not based on turnover.
OFF is UKAS accredited certification body according to EN 45011 for production and processing.

Biodynamic Agricultural Association (BDA Certification)

Offering organic and Demeter (biodynamic)certification

GB-ORG-06

Contact address

The Biodynamic Association, Open House,
Gloucester Street, Stroud

Tel: 01453 766296

Email: certification@biodynamic.org.uk

Website: www.bdcertification.org.uk

Standards: Organic and Demeter Standards for production and processing can be downloaded free from the website.

Inspection and certification charges:

Please contact the office

Soil Association Certification

GB-ORG-05

Contact address
Spear House, 51 Victoria Street, Bristol BS1 6AD
Tel: 0117 914 2406
Email: goorganic@soilassociation.org
Website: www.soilassociation.org/certification

Standards: can be downloaded from the website.
www.soilassociation.org/what-we-do/organic-standards

Inspection and certification charges* (excl.VAT):

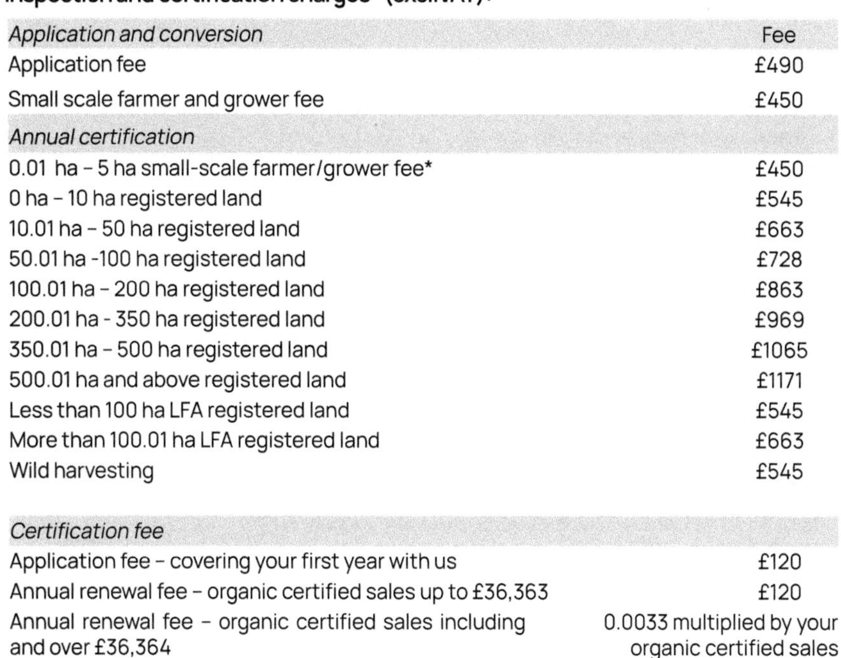

Application and conversion	Fee
Application fee	£490
Small scale farmer and grower fee	£450
Annual certification	
0.01 ha – 5 ha small-scale farmer/grower fee*	£450
0 ha – 10 ha registered land	£545
10.01 ha – 50 ha registered land	£663
50.01 ha -100 ha registered land	£728
100.01 ha – 200 ha registered land	£863
200.01 ha - 350 ha registered land	£969
350.01 ha – 500 ha registered land	£1065
500.01 ha and above registered land	£1171
Less than 100 ha LFA registered land	£545
More than 100.01 ha LFA registered land	£663
Wild harvesting	£545
Certification fee	
Application fee – covering your first year with us	£120
Annual renewal fee – organic certified sales up to £36,363	£120
Annual renewal fee – organic certified sales including and over £36,364	0.0033 multiplied by your organic certified sales

Additional inspection services, including Red Tractor, aquaculture and seaweed production and a Local Abattoir/Butcher scheme, as well as technical development opportunities such as events, workshops and programmes like as Innovative Farmers.
The Soil Association also offers certification for forestry, processor, catering, health & beauty and textiles operations – please visit the website for further details.or contact their office.

Full details of fees for all services will be sent with the initial information pack.

* These are the fees at the time of writing, they may be reviewed with each financial year.

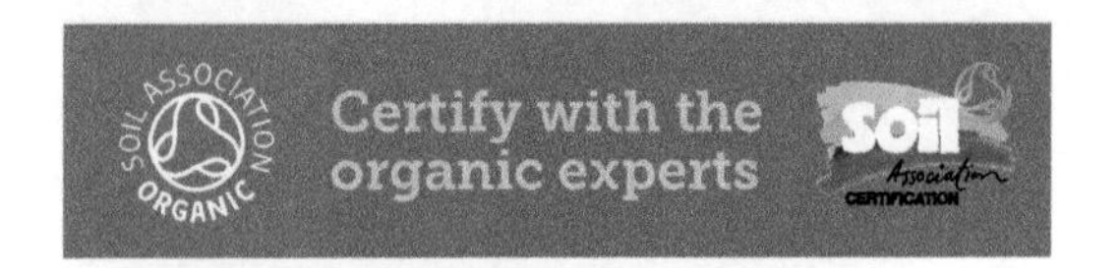

Irish Organic Association

Operates in the Republic of Ireland and in Northern Ireland

GB-ORG-07
IE-ORG-02

Contact address

Unit 13, Inish Carraig, Golden Island, Athlone, Co. Westmeath, N37 N1W4, Ireland

Tel: (+353) 090 643 368

Email: info@irishoa.ie

Website: www.irishorganicassociation.ie

Standards: IOA Standards for Organic Food and Farming in Ireland. Free.

Inspection and certification charges: Please contact the office directly.

Annual renewal fees: (including membership), please contact office.
N.B. clients may be liable for any extra costs due to follow-up inspections, etc. (more details on request).

See website for other services (e.g., training and advice resources).

SECTION 3
CERTIFICATION

Organic Trust CLG.

Operates in the Republic of Ireland and in Northern Ireland

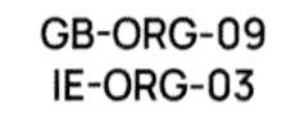

Contact address

Office A1, Town Centre House, Naas Town Centre, Dublin Road, Naas, Co. Kildare, W91 KVX8

Tel: (+353) 45 882 377

Email: info@organictrust.ie

Website: www.organictrust.ie

Standards: OTL Organic Food and Farming Standards in Ireland.Free on CD and on-line.
Full application pack (including Standards Manual).Free.

Inspection and certification charges: Please contact the office directly

Other: List of recommended organic advisers and training courses.
Magazine: Clover magazine – free to all licensees.

Quality Welsh Food Certification Ltd.

GB-ORG-13

Contact address

PO BOX 8, Gorseland, North Road, Aberystwyth, Ceredigion
SY23 3SD

Tel: 01970 636 688

Email: organic@wlbp.co.uk

Website: www.welshorganic.co.uk

Standards: Welsh Organic Scheme Standards can be downloaded from the website.

Inspection and certification charges (excl. VAT):.

LFA 0.1– 10 ha	£456
LFA 10+ ha	£490
Non LFA 10 + ha	£461
Non LFA 10.1-50 ha	£571
Non LFA 50-100 ha	£588
Non LFA 100-200 ha	£703
Non LFA 200-500 ha	£721
Non LFA 500.01 + ha	£841

Quality Welsh Food Certification is UKAS accredited certification body according to EN 45011.

The conversion process

The conversion (or transition) from conventional to organic farming systems is subject to several physical, financial and social influences which differ from those associated with established organic farming systems. During the conversion period, the aim should be to:

- improve soil fertility by establishing a rotation with legumes (e.g., clover grass leys) in order that crops can be produced without synthetic nitrogen fertiliser or large amounts of bought-in manures;
- adjust the stocking rate to the natural carrying capacity of the farm, so that livestock can be produced without high levels of purchased concentrates and/or purchased forage;
- change the management of the system to maintain animal and plant health with the limited inputs acceptable to organic production standards;
- work towards establishing a balanced ecosystem and maintaining species diversity to benefit the environment and agricultural production.

The time required and the difficulties associated with the necessary changes will depend on the intensity of conventional management and the condition of the farm before conversion, the extent to which new enterprises and marketing activities are introduced and any yield and financial penalties related specifically to the conversion process. The process can take several years and be complex, particularly when converting intensive livestock farms or cropping farms, involving innovation and restructuring of the farm system as well as changes in production methods. In other situations, conversion may be very straightforward. For example, mixed farms operating with lower agrochemical and veterinary inputs, at lower stocking rates and with good rotations are relatively straightforward to convert.

After the statutory (see Section 3) conversion period of two-years for annual crops (three years for permanent e.g. fruit crops), the land can qualify for full organic certification, and will be allowed to use the GB Certification Code (compulsory on packaged products) and/or the relevant Control Body symbol to market the products and get access to premium prices where they are available. Products from land in the first year of conversion still counts as conventional but products from the second or third years of conversion can be labelled as 'in-conversion to organic farming', providing they abide by the labelling conditions set out in the organic regulation. At no point can products be labelled as organic before the conversion period has elapsed.

Conversion-related innovation

The conversion process often involves innovation on the part of the farmer, including the introduction of:

- new marketing approaches, such as direct marketing to consumers, local shops or specialist organic retailers, as well as adding value and processing (e.g., milk into yoghurt or cheese) in order to obtain a premium;
- new crop production practices and inputs, such as mechanical weed control, biological controls, clover/grass mixtures, undersowing, intercropping and new tillage practices;
- new crop and livestock enterprises such as potatoes, grain legumes, field-scale vegetables and integration of sheep and cattle;
- new livestock management practices such as more natural rearing systems, appropriate housing, complementary animal therapies like homoeopathy and the integration of livestock into the whole farming system;
- new manure management practices such as composting and slurry aeration;
- new labour sources such as students or casual labour, or person with specific skills for production of marketing activities.

Conversion does need new techniques to be learnt and an increased level of observation and flexibility as well as management expertise and confidence from the farm staff. There is therefore a need for training and advice in preparation for and as part of the conversion process (see Sections 14 and 15 for further information).

Approaching the conversion

Interest in conversion is often triggered by an event or a change in circumstances, both on the farm or in the wider context. At first, the farmer needs to develop the confidence that organic farming will suit their farm and be prepared to find creative solutions to overcome some of the problems which may be encountered. The early stages will usually involve visits to other organic farms, attendance at seminars and conferences (see Section 14), discussions with advisers and farmers, reading available literature including what is available online. This information-gathering exercise may then be followed by a 'trial' phase. There may be some difficulties associated with the 'trial' phase. While trying organic management on one field allows direct experience to be gained, it does not allow the establishment of a suitable rotation with fertility-building crops, the integration of livestock and the management of the farm ecosystem for pest and disease control. Similarly, gradual de-intensification,

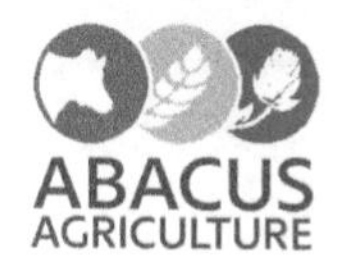

or experimenting with the withdrawal of inputs, does not provide direct experience of organic farming as a complete system. Conversion planning, aimed at assessing the feasibility of conversion for the individual farm (see below), can often replace the 'trial' phase.

Once the decision to convert the farm has been taken, all that has been learned so far will be put into practice and access to markets will need to be developed. There are two main approaches to converting a farm:

- *Staged (or phased) conversion* involves the conversion of parts of the farm, typically 10-20%, in successive years, using a fertility-building legume crop as an entry into organic management of arable land or establishing clover based swards on pasture land. The learning costs, capital investment and risks can be spread over a longer period, sometimes up to 10 years or the full length of a rotational cycle, and are more easily carried by the remainder of the farm business; also livestock enterprises usually follow after the grassland has been converted to allow for sufficient forage
- *Single-step conversion* involves converting all the land on the farm at one time. This enables the farm to gain access to premium prices sooner, but means that all the risks, learning costs and financial impacts of conversion are concentrated into a short period of time. Rotational disadvantages can arise because it may not be realistic to put all of the fields into fertility-building crops at the same time. If mistakes are made, the impacts are likely to be more severe and the approach may turn out to be more costly, despite the earlier access to premium prices.

Careful planning of the conversion should allow choosing the best strategy for a particular farm. Both staged and single-step conversions involve compliance with organic production standards on the land that is being converted from the start of the conversion period, but with some exceptions access to premium markets is generally not possible until the official two-year conversion period has been completed. Staged conversion is less suitable where it is important to gain organic status and premiums on the livestock at an early stage.

Reductions to the conversion period are only possible on land that has been under contractual agreements that exclude the use of non-permitted inputs (e.g., agri-environment schemes) and this has to be authorised in advance. It is the exception and will only be granted where there is sufficient proof that non-permitted inputs have not been applied for at least three years. Please contact Control Bodies for further details.

Conversion-specific yield reductions

There is some evidence of a decline in yield during the conversion period greater than that which would be expected in an established organic system. This is because biological processes such as nitrogen fixation and rotational effects on weeds, pests and diseases take time to become established. Yield reductions can also occur because of mistakes or inappropriate practices, such as the removal of nitrogen fertiliser without taking action at the same time to stimulate biological nitrogen fixation using legumes.

In many cases, conversion-specific yield reductions can be avoided where:

- the farmer has access to information and the opportunity to learn about organic management before starting conversion;
- the conversion is undertaken in stages and starts on each field with a fertility building forage legume, such as a clover/grass ley;
- the change over to the organic rotation for each field is carefully planned.

Grasslands are one area where conversion-specific yield decline may be inevitable, with production lost as a result of reseeding grassland with new mixtures containing legumes or waiting for clover to establish naturally following the withdrawal of nitrogen fertiliser. The loss of output can place significant pressure on stocking rates for livestock on predominantly grassland farms, particularly where permanent grassland is involved.

Changes in labour requirements[1]

Studies of organic farming indicate that increases in labour use are typically in the range 10-25%, although on livestock farms where stock numbers are reduced total labour requirements may actually fall. These increases are associated with the introduction of more labour-intensive (but high-value) crops and/or production techniques, on-farm cleaning, grading, processing and marketing of produce, small-scale experimentation with new crops as well as increases in farm size. There may be conversion-specific increases in labour requirements associated with market development, innovation, and delays in labour-saving investments such as mechanisation due to cash flow constraints, but research confirms this is limited. There is little evidence that labour requirements for existing enterprises increase during conversion, although labour use per animal may increase through preventive health management, observation and where intensive livestock enterprises are converted to free range systems.

---

1. Orsini S, Padel S, Lampkin N (2018) Labour Use on Organic Farms: a Review of Research since 2000. Organic Farming 2018 I Volume 4 I Issue 1 I Pages 7–15 DOI: 10.12924/of2018.04010007

Once converted, UK surveys of farm income indicates that there may be very little difference in labour requirements on organic and similar conventional holdings (see Section 5) except in vegetable production.

Conversion costs

The costs of conversion vary widely according to individual circumstances, as a result of a combination of one or more of the following:

- *Output reductions* due to changes in husbandry practices, enterprise mix (including a possible conversion-related emphasis on fertility-building legumes at the expense of cash-cropping), or as a result of mistakes or inappropriate actions which could be avoided through improved information and planning.
- *New investments* in land, machinery, livestock, buildings, fences, water supplies, feeding systems, manure handling systems and other facilities, in particular when converting from very specialised arable or livestock systems – some of these may be financed by capital released through reductions in livestock numbers.
- *Information and experience gathering* including direct costs, e.g., for literature, training courses, advisory services, study tours, seminars and conferences, and indirect costs, such as replacement labour to cover for absences on training courses and crop failures/poor performance where new enterprises or techniques are being tested.
- *Variable cost reductions* as prohibited inputs are withdrawn and other inputs such as purchased livestock feeds are restricted – although some additional costs may be associated with reseeding grassland, establishing green manures and other fertility-building measures.
- *Fixed cost increases*, normally restricted to labour use, depreciation of conversion-related investments, and certification charges. Higher depreciation costs may also be associated with the writing-off of investments in discontinued enterprises (such as intensive livestock housing).
- *Lack of access to premium prices*, potentially worth £200-£500/ha, during the official two-year conversion period. Where premium prices are available, significant *market development costs* may be associated with obtaining them.
- *Eligibility for conversion payments and other support schemes*, such as access to Organic Entry Level Scheme and in some cases Higher Level Scheme in England, or their replacement.

Improved information, conversion planning and marketing can help to reduce the costs of conversion and improve returns during the conversion period. There is also support through the Organic Farming Schemes (see below), but this may only partly compensate the costs, particularly where there is no access to premium prices. The choice between staged or single step conversion (see above) needs to consider the implications of this. Some producers have opted for a simultaneous conversion of all land and livestock, which requires managing stock according to organic standards from the start. Under those conditions, it might be worth considering extending the normal two-year conversion period if no market outlets have yet been identified. This is a commercial decision and needs the agreement of the Control Body.

Specialist or mainly arable farms and horticultural holdings converting to organic production will tend to reduce the arable area and increase the area of fertility-building grassland and numbers of livestock. The temptation to grow in-conversion cereal crops to generate cashflow should be balanced against the need for fertility building. Fertility building is best undertaken when the income from conversion support is greatest, early in the scheme.

Specialist livestock farms will tend to reduce livestock numbers and change the livestock enterprise mix and may introduce or expand arable production. Dairy producers are more likely to try to maintain cow numbers by increasing the forage area available to them and therefore reduce other enterprises or increase farm size. Mixed arable/livestock farms are less likely to need structural change of this type.

Conversion planning

Conversion planning with the assistance of specialist advisers can have several purposes:

- Provide a feasibility assessment about what changes conversion of a farm to organic would involve and support the decision on whether or not to go ahead with conversion.
- Guide through the conversion process itself, i.e., planning a rotation and cropping plan, identify potential problems relating to labour requirements, other resource constraints, forecast financial returns and cashflow and that help reduce the risk.
- Support the applications for certification and for conversion support.

Detailed conversion planning involves:

- an assessment of the current farming system and management including the farmers' objectives for the conversion and organic farming is an important element of the conversion planning process which provides the basis for identifying personal and resource limitations that might be faced during the conversion;
- development of proposals for a 'target' organic endpoint, which require the design of an appropriate rotation and stocking plan, in line with the farmer's objectives and marketing options and the testing of the technical and financial feasibility of the proposed production activities;
- a strategy for getting from the current situation to the target including the conversion strategy (i.e. staged or single-step conversion), as well as the timing of changes in the management and extent of crop and livestock enterprises, major soil improvements, changes in land and labour resources and capital investments.

A detailed plan should include steps to enhance animal health and welfare (see Section 10) as well as the environmental and nature conservation characteristics of the farm (see Section 13). The financial implications, in particular cashflow and tax implications during the transition phase, also need to be considered. A full conversion plan should therefore involve analysis of both financial and physical aspects of the whole farm business. The plan must take account of the long-term effects on cash flow and the farm system and allow re-planning during the conversion period to accommodate any change.

The cost of advice for a conversion plan varies according to the size, type and complexity of the farm and the level of detail. For a medium-sized family farm, a basic service including 2 or 3 farm visits by a trained adviser (see below), summarising the key management changes and cropping programme, will cost between £800 and £2,000.

More restricted conversion plans, not usually requiring financial details, are a requirement for certification and some organic support schemes.

Organic conversion information and advisory services

General information on conversion and related certification issues can be obtained from organic organisations and Control Bodies (see Sections 3, 14 and 15).

Abacus Agriculture
Organic farming
advice that counts...

Control Body Organic Farmers and Growers Ltd and consultants Abacus Agriculture offer a free advisory service for producers interested in organic conversion https://organicinfo.org.uk

Control Body Soil Association Farming and the Land Use team offer advice www.soilassociation.org/farmers-growers/meet-the-team/

Contact details of advisory organisations providing organic advice are provided in Section 15.

On-line resources relating to organic conversion include:

- Agricology: www.agricology.co.uk
- Gov.uk Organic Farming:
  www.gov.uk/guidance/organic-farming-how-to-get-certification-and-apply-for-funding
- Ireland: Teagasc
  www.teagasc.ie/rural-economy/organics/steps-to-organic-conversion
- Northern Ireland: DAERA/CAFRE
  www.daera-ni.gov.uk/topics/livestock-farming/organic-farming
- Scotland: SRUC/SAC Consulting
  www.sruc.ac.uk/media/uuchplwc/tn520-convert-organic.pdf
- Wales: Welsh Government
  www.sruc.ac.uk/media/uuchplwc/tn520-convert-organic.pdf

Some specific technical guides on conversion are available, for example:

- Converting to organic farming Organic Research Centre
  www.organicresearchcentre.com/farming-organically/converting-to-organic-farming/
- Producer Conversion Plan Organic Farmers and Growers Ltd.
  https://assets.ofgorganic.org/rd92-producer-conversion-plan.9hh63v.pdf
- Organic arable farming: information for farmers considering conversion to organic production. Organic Farmers & Growers Producer Technical Leaflet TL 128. https://assets.ofgorganic.org/tl-128-organic-arable-farming.ipi4sm.pdf
- Guide to Certification – Farming Soil Association
  www.soilassociation.org/media/23822/guide-to-organic-certification-farming.pdf

Organic producer and membership organisations also provide support during conversion.

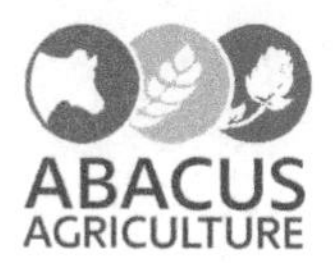

In **Scotland**, SRUC provide advisory services and online information services. See:

- https://tinyurl.com/SAC-consult-organic

The Scottish Government Veterinary and Advisory Services (VAS) programme provides animal health information. SAC Consulting is a division of SRUC (Scotland's Rural College). For further information:

- www.sruc.ac.uk/business-services/sac-consulting/contact-sac-consulting/

In **Wales**, Farming Connect offers organic business advice:

- https://tinyurl.com/FC-organic

In **Northern Ireland**, the Department of Agriculture, Environment and Rural Affairs offers information on organic farming and links to further support via its website:

- https://www.daera-ni.gov.uk/topics/livestock-farming/organic-farming

In the **Republic of Ireland**, a 25 hour *Introduction to Organic Production* course must be completed before acceptance into the Organic Farming Scheme (see below). For details of courses held nationwide see:

- www.teagasc.ie/rural-economy/organics

Conversion support under the organic farming grant schemes

Support for organic conversion and maintenance in most parts of the UK (not all). Typically, conversion payments are for the first 2 years (and in some cases depending on crop – perennial fruit is 3 years) and will be at higher rates than organic maintenance funding. At the time of print, the application window for 2023 for England is closed with newly awarded payments to commence on the 1st of January 2024. Potential applicants should check their relevant administration's websites for updates. Schemes are normally subject to application windows, five-year agreements (including some period as part of maintenance payments) and certification by an approved body and may involve compulsory or optional combinations with other agri-environment options (see Section 13). Unlike other agri-environment schemes, applicants for the organic support options must be 'active farmers'.

Since the UK-exit from the European Union, payments for organic conversion and maintenance support are going through a period of transition as new agricultural funding structures are introduced in all four of the devolved

Abacus Agriculture
Organic farming
advice that counts...

administrations (DAs). This has led to considerable uncertainty about future funding and many of the suggested payment rates in this section are based on 2023 rates, and 2024 rates are as yet not confirmed. Any relevant updates on these will be provided in the Organic Research Centre e-bulletins[2] and on the Organic Farm Management Handbook updates page on the ORC website (www.organicresearchcentre.com).

England Countryside Stewardship (CS)

Financial support for conversion to organic farming is available in the 2023 Mid and Higher Tier elements of Countryside Stewardship (CS), with a range of organic conversion and organic land management grants and options.

Payments under the scheme are available on land registered as 'in conversion' with a Defra-licenced organic control body (CB) across five different farm types for a maximum of 2-3 years. A conversion plan should be agreed with the CB and supplied alongside each application. Land should be registered as fully organic by the end of the agreement and farms move on to related Mid-Tier organic land management grants (see Section 5). Some conversion options also permit the application of other Countryside Stewardship options on the same area of land. Conversion grants are not available for land that has previously been registered with an organic control body by the applicant or land for which an applicant has previously received conversion aid. For full details on criteria, applications and future funding rounds see:

- www.gov.uk/guidance/organic-farming-how-to-get-certification-and-apply-for-funding

Since the UK has left the EU, Countryside Stewardship now forms part of the new funding structure: Environmental Land Management (ELMS). Many features of previous schemes have been retained. Two new schemes are also in the process of being introduced and are expected to be fully operational by 2025. The Sustainable Farming Incentive (SFI) will replace the Basic Payment Scheme and provide funding opportunities for various sustainable farming practices. There will be the option for organic farmers to stack some SFI payments on top of their organic conversion and/or maintenance funding, see Section 13. The other new funding scheme, Landscape Recovery, targets larger scale collaborative projects to enhance the natural environment. Contracts under Landscape Recovery will be long term once awarded. The scheme is expected to be implemented in 2024 and is currently being piloted in collaboration with the North East Cotswold Farm Cluster[3].

---

2. Subscribe at: www.organicresearchcentre.com/e-bulletin/
3. www.cotswoldfarmers.org/
https://agricology.co.uk/research-projects/north-east-cotswold-farmer-cluster-cic/

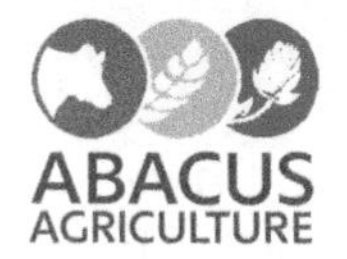

The payment rates for Organic Conversion grants are as follows:

English CS Organic Conversion grants payment 2023 (£/ha & period)

	£/ha per year	Notes
Improved permanent grassland (OR1)	187	Arable land reverted to permanent grassland. Whole parcel for 2 years.
Unimproved permanent grassland (OR2)	89	Unimproved permanent grassland and rough grazing below the moorland line. Keep whole parcel for 2 years.
Rotational land (OR3)	296	Arable land, temporary grassland or permanent pasture cultivated in preceding 7 years. Whole or part parcel for 2 years.
Horticulture (OR4)	703	Arable land, temporary or permanent grassland where cultivation is planned as part of conversion. Max 20 ha. Whole or part parcel for 2 years.
Top Fruit (OR5)	1254	Top fruit and permanent bush crops. Orchards used in production of alcoholic drinks or not in commercial production are not eligible. Whole parcel for 3 years.

Scotland: Agri-Environment Climate Scheme

The Scottish organic conversion support scheme is funded through the Scottish RDP up to and including 2024, with provisions for support for organic conversion and maintenance available through the Agri-Environment Climate Scheme (AECS). Scottish Ministers have committed to existing Scottish payment schemes for organic conversion and maintenance until 2024, but as yet there has been no official confirmation provided as to how organic payments will be delivered under the new Agricultural Bill. It is assumed however that there will be funding opportunities for organic agriculture under the forthcoming band of 'enhanced conditionality' payments.

Applications are administered through the Scottish Government's Rural Payments and Services. The 2023 application round was opened between 30 January 2023 and 7 June 2023. Awarded applications for this period will then receive funding in 2024. Organic maintenance and conversion payments are no longer capped in AECS for the 2023 application round. Existing organic maintenance contract holders with land on their schedule of works in excess of the previous caps will start to be paid on these areas from the 2024 scheme year.

Details for the 2024 application window were not available at the time of writing. For further information on the 2024 scheme and latest developments, see: https://tinyurl.com/Scot-conversion-pay

Abacus Agriculture
Organic farming
advice that counts...

Scotland Organic Conversion Payment rates 2023 (£/ha & year)

Land type*	Years 1-2	Years 3-5	Total (5 years)
Arable	280	65	755
Vegetables and fruit	400	200	1400
Improved grassland	140	55	445
Rough grazing	12.50	8.50	50.5

** Minimum payment of £500 per agreement*

Financial support for capital items associated with conversion to organic production (e.g., fencing costs) was available in 2023 if ample justification is provided. Organic farmers would not usually be eligible for this payment, however in cases where (for example) additional costs are clearly present as a result to converting to an organic rotation from (for example), continually cropped or continually grazed land.

Wales: Glastir Organic

Funding for conversion to organic farming has not been available in Wales since 2018, except for a one-off scheme in 2022 providing funding in 2023 and 2024. The Sustainable Farming Scheme will be introduced in 2025. (see Section 13 for more relevant information).

Under the 2022 Organic Conversion Scheme, support for land under conversion was set at four payment rates (see Table). Payments are capped according to the following:

- 0 - 200 ha of eligible land: 100% of the payment rate
- 200 – 400 ha of eligible land: 50% of payment rate
- 400 ha+: 10% of payment rate

2022 Organic Conversion Scheme payment rates

	£/ha per year	Notes
Payment Rate 1 (rotational land)	202	Land within a rotation, such as arable land, grassland within an arable rotation, temporary grassland or horticultural crops within a rotation, such as brassicas, potatoes.
Payment Rate 2 (permanent crops/ grassland)	101	Land not within a crop rotation, permanent crops, permanent grassland. Permanent horticultural crops, such as orchards, soft fruit etc.
Payment Rate 3 (Permanent and temporary grassland with a dairy enterprise)	345	Land not within a crop rotation and permanent grassland. Temporary grassland. Farms with multiple enterprises, for example, a dairy unit and upland sheep will only be paid payment rate 3 for land utilised by the dairy enterprise. Land utilised by the sheep enterprise will be paid on either payment rate 1, 2 or 4.
Payment Rate 4 (unenclosed land)	12.6	Applications for this rate will include unenclosed upland and sole grazier, grazed common land field parcels.

** A contribution towards certification costs of £500 per year for the 2 year conversion period will be provided.*

Northern Ireland: Organic Farming Scheme (OFS)

The Department of Agriculture, Environment and Rural Affairs formerly provided grant funding for conversion under the Environmental Farming Scheme, which is currently closed for 2023. See:

- https://tinyurl.com/DAERA-conversion
- Tel. 0300 200 7841

Republic of Ireland: Conversion support

Organic conversion payments are administered through Ireland's Organic Farming Scheme (OFS) program. At the time of writing, this is no longer accepting new applications for 2023. Funding for the Organic Farming Scheme is in place until 2027. Under the OFS both existing and converting organic producers were supported. In-conversion payment rates vary according to size and land use and are only paid for the first two years, after which time the lower organic rates (see Section 5) apply. The rates are as follows:

Abacus Agriculture
Organic farming
advice that counts...

Irish conversion payment rates 2023

	Years 1-2 (in conversion rates 1-70ha) €/ha	Years 1-2 (in conversion rates >70ha) €/ha	Notes
Horticulture	800	60	Organic horticulture producers, with an organic horticulture area of one hectare or more, are eligible provided that at least 50% of the area eligible for organic payment is cropped each year.
Tillage	320	60	Organic tillage producers, with an organic tillage area of six hectares or more.
Dairy	350	60	Dairy producers, with an organic area of six hectares or more.
Dry stock (beef & sheep) and all other holdings	300	60	Applicants with 3 hectares or more of utilisable organic agricultural area are eligible.

An annual participation payment will be paid to all OFS participants to cover administrative costs in addition to the per hectare payments detailed above. An amount of €2,000 will be paid to OFS participants in the first year of conversion and €1,400 for every subsequent year of the contract.

More information regarding organic conversion in Ireland can be found at the following address:

- https://tinyurl.com/Teagasc-conversion

Step by step guide to organic conversion

1 **Find out more about organic farming**

- Visit organic farms to see them working
- Attend events and read more about it
- Read the organic standards

2 **Evaluate organic farming for your farm in more detail**

- Assess your farm for its suitability for conversion, possible with help of an advisor
- Consider whether organic farming suits the personal ambitions of you and your family
- Assess the organic market to see if there are wholesale or retail opportunities for the potential organic products of your farm
- Investigate grant opportunities
- Identify a suitable Control Body, talk to them and get an application pack
- Possibly some practical experiments, e.g., clover grass sward on a small number of paddocks, trying different weed control techniques, introducing cover crops etc.

3 **Draw up a detailed conversion plan**

The conversion plan should include:

- Assessment of the farm including current farming system, infrastructure, weed levels, soil type, skills
- Description of the proposed organic farm system including crop rotation, livestock numbers and type, soil, stock and crop management, labour, buildings and machinery requirements
- Timescale for grant and certification application, field by field and livestock conversion dates (start and completion), capital investment, training needs (including of people working with you)
- Cash flow, investment and Profit/Loss budget

4 **Apply for organic conversion/certification and organic grants**

5 **Start implementing organic management on the farm according to the plan**

6 **Review and revise**

- Be prepared to revise the conversion plan regularly in the light of your experience, but maintain basic needs for fertility building and management of crop and animal health

An experienced organic adviser can help throughout this process, providing technical, marketing and business advice.

Why a whole-farm approach?

Organic farming is best approached on a whole farm basis, not as a diversification opportunity for individual enterprises as part of a predominantly non-organic holding. Although UK organic regulations provide for certification of part farms, these are required to be operated as independent units financially and managerially. There are restrictions on the parallel production of organic and non-organic crops and livestock. More fundamentally, as detailed in Section 1, many of the husbandry practices that make organic farming work effectively, such as crop rotation and clean grazing systems for parasite control, can only be achieved where the management of different enterprises is integrated in a whole farm approach. This is captured in the idea of the farm as an organism or a system, where the different elements – soil, plants, animals, people and other resources/activities – interact to create a self-governing, stable and resilient whole that is greater than the sum of its parts. Like an organism, and consistent with the ecological principles behind organic farming, the system should be designed in such a way that it can resist and recover from external shocks and pressures, such as disease, drought or economic recession. This is helped where the outputs from one activity or element of the system act as inputs for others and help sustain a productive farming system.

This section looks at the financial performance of organic farming from a whole farm perspective, by looking at projected 2023 gross margin results for different systems based on the financial data contained in the later sections and by looking at the financial performance of actual organic farms in 2021/22 in England and finally by considering the role of whole farm support payments including organic farming maintenance support and some options of the Sustainable Farming Incentive that might be relevant.

Whole-farm financial performance

Whole-farm gross margins

A comparison of the individual enterprise gross margins presented in Sections 7-11 with conventional ones indicates that for most crops, the combination of lower yields, lower variable costs and higher prices can lead to higher gross margins than in comparable conventional systems. For livestock (except dairy), at times a lower level of premium prices meant that although similar performance per animal could be achieved, gross margins per animal tend to be lower. This is also true per hectare

due to reduced stocking rates. Premiums for organic livestock have been variable and gross margins are also strongly influenced by feed prices.

In organic systems especially, it is necessary to consider individual enterprises in terms of their contribution to the whole farm system. High gross margins for individual crops need to be seen in the context of the rotational constraints, which, for example, prevent wheat being produced as often in the cropping sequence as might be possible in conventional systems. In contrast, ruminant livestock enterprises represent an opportunity to utilise fertility buildings crops that have no direct return as such. Depending on the combination of crop and livestock enterprises, whole farm gross margins may be higher or lower than in conventional systems.

The following 'thumb-nail' sketches and gross margin models for different types of organic farming system illustrate some of the issues which must be considered when looking at the viability of the whole organic business. While they each show different rotational constraints and opportunities, they are not the only possible rotational options. It is important to seek professional advice on rotations appropriate to specific individual circumstances.

The systems shown are the same as in previous editions of the Organic Farm Management Handbook. Fertility building leys can be used by livestock or for seed production or cut and mulched as required. The whole farm gross margins for earlier years are shown for comparative purposes as a matter of interest only. They are not a reflection of trends recorded on individual organic farms, but they illustrate the effects that changes in the gross margins might have on the overall financial results of farms.

Organic farm income survey

Until 2022, Rural Business Research based at various universities carried out data collection for the Farm Business Survey (FBS) in England. This included annual reporting on the financial performance of organic farms in England[1], but sample sizes for several farm types have fallen in recent years, making the data less robust. From 2023, the Farm Business Survey in England will be provided by a different contractor, and it is not clear whether comparisons between organic and conventional samples will be provided and on what basis comparisons might be undertaken. Previously, Aberystwyth University and ORC carried out Defra-funded farm business monitoring of organic farms of different types in England and Wales. A report looking at six-year trends for this data (covering the period 2005/06-2011/12) was produced for the Soil Association by ORC[2] .

1. See publications at https://www.ruralbusinessresearch.co.uk/
2. Lampkin, N; Gerrard, C; Moakes, S (2014) Long term trends in the financial performance of organic farms in England & Wales, 2006/07-2011/12. Report for Soil Association by Organic Research Centre

The results presented here are the average for the group of organic farms of a particular type, compared with the average of similar conventional farm types. The 2021/22 report contains results for an identical sample of organic farms of each type in two consecutive years in England. For some farm types these are less than 15 observations in the sample which could reduce the robustness of the results, due to the impact that dominant individual farms can have on the group averages. It has not been possible to publish data for specialist pig and poultry farms, due to small sample sizes.

Comparing 2021/22 results with the same farms in the previous year (see Table), the overall Farm Business Income (FBI) per ha on organic horticulture and LFA cattle/sheep farm types fell, while cropping, mixed, dairy and lowland cattle/sheep incomes increased, despite the emerging challenges at the end of 2021. Most organic farm types achieved lower incomes per ha than comparable conventional farms, although organic lowland and LFA cattle and sheep performed nearly as well as conventional farms. It should be noted that the results refer only to England and both the sample sizes (several farm types with less than 15 farms) and the average farm sizes have reduced significantly compared with 2014/15. This indicates that the results are less reliable than in previous years.

Comparison of organic Farm Business Income trends by Farm Type, 2021/22 and 2020/21 (identical samples)*

		2021/22		2020/21	
Farm Type	**Farms**	**£/farm**	**£/ha**	**£/farm**	**£/ha**
Cropping	7	39468	367	29886	349
Horticulture	7	19957	1323	35877	2674
Mixed	8	18850	118	17171	105
Dairy	29	53479	315	50963	292
Lowland cattle and sheep	31	26622	327	17135	218
LFA cattle and sheep	18	46105	275	42997	325

** A part of this table was derived from data with less than 15 observations in the sample which could reduce the robustness of the results*

Source: Scott C (2023) Farm Business Survey 2021/2022 Organic Farming in England. Rural Business Research at Newcastle University, Newcastle.

Decisions on conversion to or remaining organic need to consider that results from survey data are not necessarily directly applicable to individual farm situations. Also, organic farming represents a long-term commitment due to the two or three-year conversion period and the time it can take for the organic

system to become fully established. It is important to take account of the long-term trends in profitability and avoid taking short-term decisions during periods of fluctuating markets.

Although now dated, the long-term trend comparisons produced by ORC based on larger sample sizes and careful selection of comparable conventional farms indicates that the financial performance of organic and non-organic farms of different types has been quite consistent. The results for horticulture should be treated with caution due to the small sample size and possible poor matching of farms between the two samples with a variety of holdings in this category (including glasshouses, market gardens, field vegetable and fruit production).

In the following pages, detailed results for the different farm types are presented, featuring both 2023 whole farm gross margin calculations, based on the data presented in this handbook, and the 2021/2022 survey results.

In the model calculations we have shown an illustrative SFI payment of £200/ha that most organic farmers should be able to achieve (see details below).

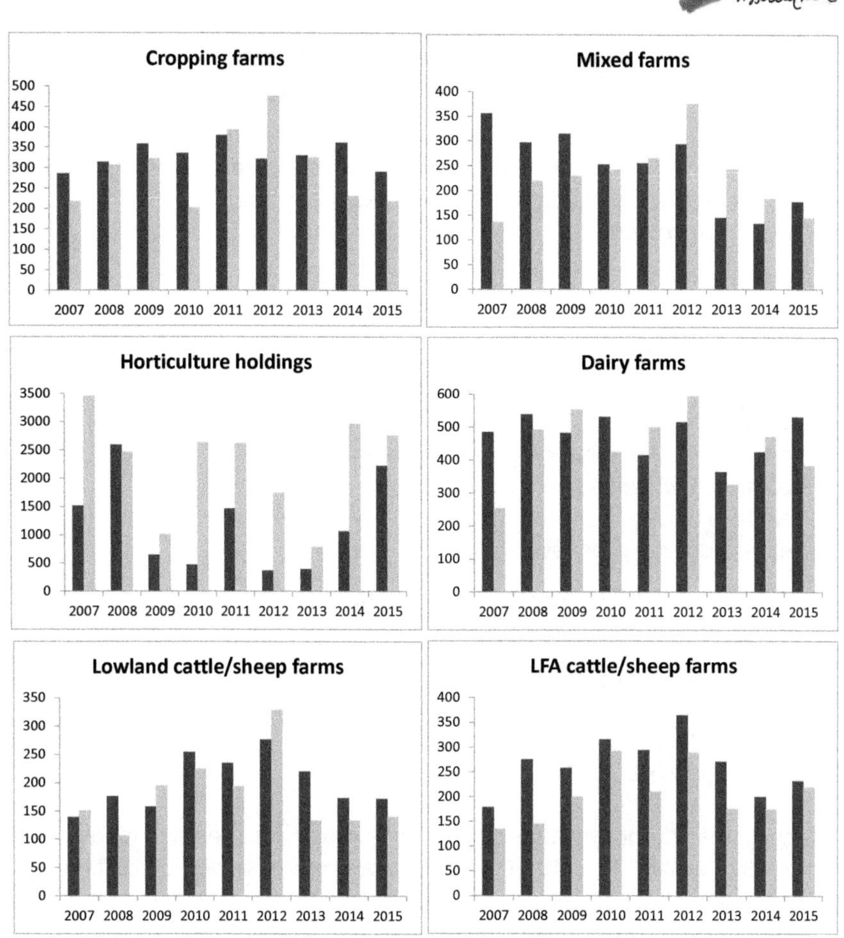

SECTION 5
WHOLE FARM PERSPECTIVES

Long term Farm Business Income trends (£/ha) on surveyed organic and comparable conventional farms, 2006/07-2014/15, full (not identical) samples

Arable systems

In organic arable systems, less reliance is placed on medium to long-term leys grazed by livestock for the fertility-building phase of the rotation. Instead, grain legumes (e.g. field beans) and legume-based green manures (e.g. red clover mixtures) can be used.

Gross margin models

The mainly arable system is based on a rotation of:

3 years grass/white clover ley
1 year winter wheat
1 year winter oats
2 years red clover/Italian ryegrass ley/green manure
1 year winter wheat/potatoes/carrots
1 year field beans
1 year spring wheat/barley

The stockless arable system is based on a rotation consisting of:

2 years red clover-based green manure
1 year winter wheat/potatoes
1 year winter oats
1 year field beans
1 year spring wheat/barley

Rotations that do not include a phase of pasture for livestock feeding are technically feasible. The green manure 'crops' are managed by cutting and mulching, and the nutrients made available directly to the following crops rather than being cycled via animal manures.

For stockless systems, the income from direct payments reduces the need to rely on livestock for a financial return to the fertility-building phase of the rotation, and avoids significant capital investments in livestock, fencing, water supplies and livestock housing when converting all-arable farms.

Field-scale vegetables can be included to boost financial returns, but the extent of these enterprises is likely to be limited by available labour. The choice of potatoes and carrots is purely illustrative; leeks, onions and cabbages may also be appropriate depending on expertise and market access.

Mainly arable gross margin model				*2023 estimates*		
	Area/numbers			*Gross margin*		
Arable crops	ha	*(acres)*		£/ha	*(£/ac)*	£/farm
Winter wheat	40	*(99)*		1007	*(408)*	40280
Spring wheat	15	*(37)*		680	*(275)*	10193
Winter oats	23	*(57)*		882	*(357)*	20275
Spring barley	8	*(20)*		592	*(240)*	4736
Spring beans	23	*(57)*		898	*(363)*	20654
Potatoes, maincrop	4	*(10)*		3590	*(1453)*	14359
Carrots	2	*(5)*		3227	*(1306)*	6453
Total arable crops	115	*(284)*		1017	*(412)*	1E+05
Forage crops	ha	*(acres)*		£/ha	*(£/ac)*	£/farm
Short term leys	46	*(114)*		-181	-*(73)*	-8303
Medium/long leys	69	*(170)*		-146	-*(59)*	-10040
Permanent grassland	20	*(49)*		-90	-*(36)*	-1790
Total forage crops	135	*(334)*		-149	-*(60)*	-20133
Livestock	head	LU		£/hd	£/LU	£/farm
Beef finishing, 18m	58	24		564	1375	32705
Single suckler cows	61	74		504	413	30745
Lowland sheep	600	101		87	517	52127
Total livestock		199			581	115576
Per forage ha		1,47				856
Farm summary	ha	*(acres)*		£/farm ha	*(£/ac)*	£/farm
Arable crops	115	*(284)*	46%	468	*(189)*	116950
Forage/livestock	135	*(334)*	54%	382	*(155)*	95444
Whole farm 2023	250	*(618)*		850	*(344)*	212393
Whole farm 2017	250	*(618)*		720	*(291)*	179989
Whole farm 2014	250	*(618)*		682	*(276)*	170396
Organic Maintenance England				123	*(50)*	30760
Sustainable Farming Incentive Option England est.				200	*(81)*	23000

Stockless arable gross margin model				*2023 estimates*		
	Area/numbers			*Gross margin*		
Arable crops	ha	*(acres)*		£/ha	*(£/ac)*	£/farm
Winter wheat	30	*(74)*		1007	*(408)*	30210
Spring wheat	30	*(74)*		680	*(275)*	20385
Winter oats	40	*(99)*		882	*(357)*	35260
Spring barley	10	*(25)*		592	*(240)*	5920
Spring beans	40	*(99)*		898	*(363)*	35920
Red clover GM	80	*(198)*	33%	-296	-*(120)*	-23656
Potatoes, maincrop	10	*(25)*		3590	*(1453)*	35898
Farm summary	ha	*(acres)*		£/farm ha	*(£/ac)*	£/farm
Whole farm 2023	240	*(593)*		583	*(236)*	139937
Whole farm 2017	240	*(593)*		495	*(200)*	118852
Whole farm 2014	240	*(593)*		426	*(172)*	102215
Organic Maintenance England				132	*(53)*	31680
Sustainable Farming Incentive Option England est.				200	*(81)*	48000

Farm business survey results for cropping farms in England, 2021/22*

	Organic			Conventional		
Farm Nos./Size (ha/acres)	*9*	*112*	*(276)*	*462*	*227*	*(560)*
Financial performance	*£/farm*	*£/farm ha*	*(£/ac)*	*£/farm*	*£/farm ha*	*(£/ac)*
Crop outputs	96436	863	(349)	291925	1288	(521)
Crop inputs	20350	182	(74)	87843	387	(157)
Crop gross margin	*76086*	*681*	*(276)*	*204082*	*900*	*(364)*
Livestock outputs	289	3	(1)	9871	44	(18)
Livestock inputs	496	4	(2)	4856	21	(9)
Livestock gross margin	*-207*	*-2*	*-(1)*	*5015*	*22*	*(9)*
Whole farm gross margin	*75879*	*679*	*(275)*	*209097*	*922*	*(373)*
Other outputs	34875	312	(126)	70935	313	(127)
Agri-environmental support	24310	218	(88)	9562	42	(17)
Basic payment scheme	24050	215	(87)	43137	190	(77)
Total GM incl. subsidies	*159114*	*1424*	*(576)*	*332731*	*1468*	*(594)*
Labour	24565	220	(89)	31583	139	(56)
Contract	14157	127	(51)	27972	123	(50)
Machinery	19542	175	(71)	57432	253	(103)
General	56780	508	(206)	67234	297	(120)
Paid rent	8447	76	(31)	20597	91	(37)
Fixed costs	*123491*	*1106*	*(447)*	*204818*	*903*	*(366)*
Farm Business Income	*35623*	*319*	*(129)*	*127913*	*564*	*(228)*

**A part of this table was derived from data with less than 15 observations in the sample which could reduce the robustness of the results*
Source: Farm Business Survey data from Rural Business Research

The Farm Business Income (FBI) per ha for the organic sample in 2021/22 was 43% lower than conventional on a per ha basis. The organic farms had higher receipts from agri-environmental (including organic) support, but lower crop and livestock gross margins, as well as higher labour and general overhead costs, which may reflect the much smaller farm size of the organic sample.

Farm business survey results for mixed farms in England, 2021/2022*

		Organic			*Conventional*	
Farm Nos./Size (ha/acres)	*13*	*140.6*	*(347)*	*139*	*183.1*	*(452)*
Stocking rate (LU/farm ha)	*61.3*	*0.44*	*(0.18)*	*83.8*	*0.46*	*(0.19)*
Financial performance	*£/farm*	*£/farm ha*	*(£/ac)*	*£/farm*	*£/farm ha*	*(£/ac)*
Crop outputs	58649	417	*(169)*	152123	831	(336)
Crop inputs	18163	129	*(52)*	50000	273	(111)
Crop gross margin	*40486*	*288*	*(117)*	*102123*	*558*	*(226)*
Livestock outputs	58482	416	*(168)*	133550	729	(295)
Livestock inputs	20635	147	*(59)*	79036	432	(175)
Livestock gross margin	*37847*	*269*	*(109)*	*54514*	*298*	*(120)*
Whole farm gross margin	*78333*	*557*	*(225)*	*156637*	*855*	*(346)*
Other outputs	51144	364	*(147)*	52745	288	(117)
Agri-environmental support	19641	140	*(57)*	8728	48	(19)
Basic payment scheme	29730	211	*(86)*	35891	196	(79)
Total GM incl. subsidies	*178848*	*1272*	*(515)*	*254001*	*1387*	*(561)*
Labour	29615	211	*(85)*	25148	137	(56)
Contract	11545	82	*(33)*	16692	91	(37)
Machinery	29648	211	*(85)*	56028	306	(124)
General	68367	486	*(197)*	64960	355	(144)
Paid rent	14406	102	*(41)*	16916	92	(37)
Fixed costs	*153581*	*1092*	*(442)*	*179744*	*982*	*(397)*
Farm Business Income	*25267*	*180*	*(73)*	*74257*	*406*	*(164)*

*A part of this table was derived from data with less than 15 observations in the sample which could reduce the robustness of the results
Source: Farm Business Survey data from Rural Business Research

The surveyed organic mixed cropping/livestock farms had similar stocking rates to conventional farms. The crop gross margin was substantially lower and the livestock GM somewhat lower than conventional, resulting in a lower whole farm GM per ha. Like the cropping farms, the organic mixed farms had higher labour and general fixed costs, which may be a result of the lower farm size in a small organic sample. Farm Business Income per hectare was substantially lower than the conventional sample, despite higher agri-environmental (including organic) support payment receipts.

Dairy systems

In all organic systems involving livestock, the ley and permanent pastures must include legumes; normally a clover suited to local conditions. Livestock will be outside during the grazing season but normally housed during winter. During the period of active growth, sufficient fodder must be conserved for all the winter feed requirements of the livestock, including provision for delayed turn-out due to slower spring growth in organic systems.

Gross margin models

Specialist organic dairy farms can consist of 100% grassland, but in some cases an area of cereals, grain legumes and/or fodder row crops may be grown for use as livestock feed. In the example illustrated, the rotation consists of:

6 years grass/white clover ley
1 year winter triticale
2 years red clover/Italian ryegrass (for forage conservation)
1 year forage maize

Mainly dairy and other mixed livestock/arable farms need to achieve a balance between crop and livestock enterprises and these are often closely integrated. Rotations are often based on traditional ley farming/alternate husbandry systems. Commonly, no more than half the whole farm will be in arable crops, and it is usual to see 70-80% down to grass with 20-30% in arable crops. In this case, the assumed rotation is:

4 years grass/white clover ley
1 year winter wheat
1 year winter oats
2 years red clover ley
1 year potatoes (20%), carrots (20%), forage rye/kale (60%)
1 year spring beans/barley

The Farm Business Survey results for dairy farms in 2021/22 show that organic farms had lower stocking rates and yields per cow than conventional. Normally these differences would be compensated by higher milk prices, but the organic milk price was only a little over 5 ppl, reflecting in part the increasing prices for conventional milk at the end of 2021 as higher energy and fertiliser prices started to bite. A substantially higher premium than this is required to compensate for the reduced output on organic farms, despite the lower crop (fertiliser) costs, lower fixed costs and higher agri-environmental (including organic) support.

Specialist dairy gross margin model							2023 estimates
	Area/numbers			Gross margin			
Arable crops	ha	(acres)		£/ha	(£/ac)	£/farm	
Winter triticale	10	(25)		549	(222)	5490	
Total arable crops	10	(25)		549	(222)	5490	
Forage crops	ha	(acres)		£/ha	(£/ac)	£/farm	
Short term leys	20	(49)		-181	-(73)	-3610	
Medium/long term leys	60	(148)		-111	-(45)	-6630	
Forage maize	10	(25)		-326	-(132)	-3255	
Total forage crops	90	(222)		-150	-(61)	-13495	
Livestock	head	LU		£/hd	£/LU	£/farm	
Dairy cows (F/H)	106	106		1521	1521	161249	
Dairy replacements	21	25		1166	980	24718	
Tack sheep	100	4		12	300	1200	
Total livestock		135			1384	187167	
Per forage ha		1,5				2080	
Farm summary	ha	(acres)		£/farm ha	(£/ac)	£/farm	
Arable crops	10	(25)	10%	55	(22)	5490	
Forage/livestock	90	(222)	90%	1737	(703)	173672	
Whole farm 2023	100	(247)		1792	(725)	179162	
Whole farm 2017	100	(247)		1791	(725)	179117	
Whole farm 2014	100	(247)		1745	(4312)	2E+05	
Organic Maintenance England				132	(53)	13200	
Sustainable Farming Incentive Option England est.				200	(81)	20000	

Mainly dairy gross margin model							2023 estimates
	Area/numbers			Gross margin			
Arable crops	ha	(acres)		£/ha	(£/ac)	£/farm	
Winter wheat	10	(25)		1007	(408)	10070	
Winter oats	10	(25)		882	(357)	8815	
Spring barley	5	(12)		592	(240)	2960	
Spring beans	5	(12)		898	(363)	4490	
Potatoes, maincrop	2	(5)		3590	(1453)	7180	
Carrots	2	(5)		3227	(1306)	6453	
Total arable crops	34	(84)		1176	(476)	39968	
Forage crops	ha	(acres)		£/ha	(£/ac)	£/farm	
Short term leys	20	(49)		-181	-(73)	-3610	
Medium/long term leys	40	(99)		-118	-(48)	-4700	
Kale/forage rye	6	(15)		-359	-(145)	-2152	
Permanent grassland	10	(25)		-90	-(36)	-895	
Total forage crops	76	(188)		-149	-(60)	-11357	
Livestock	head	LU		£/hd	£/LU	£/farm	
Dairy cows (F/H)	92	92		1521	1521	139952	
Dairy replacements	18	22		1166	980	21453	
Total livestock		114			1417	161406	
Per forage ha		1,5				2124	
Farm summary	ha	(acres)		£/farm ha	(£/ac)	£/farm	
Arable crops	34	(84)	31%	363	(147)	39968	
Forage/livestock	76	(188)	69%	1364	(552)	150048	
Whole farm 2023	110	(272)		1727	(699)	190016	
Whole farm 2017	110	(272)		1688	(683)	185632	
Whole farm 2014	110	(272)		1589	(529)	174843	
Organic Maintenance England				122	(49)	13400	
Sustainable Farming Incentive Option England est.				200	(81)	22000	

Farm business survey results for dairy farms in England, 2021/22

	Organic			*Conventional*		
Farm Nos./Size (ha/acres)	*31*	*171,2*	*(423)*	*174*	*161,7*	*(400)*
Dairy cows	154			203		
Milk yield/cow, price (ppl)	5997	litres	38,2	8434	litres	32,9
Stocking rate (LU/farm ha)	*231,3*	*1,35*	*(0,55)*	*304,1*	*1,88*	*(0,76)*
Financial performance	*£/farm*	*£/farm ha*	*(£/ac)*	*£/farm*	*£/farm ha*	*(£/ac)*
Crop outputs	14804	86	*(35)*	38808	240	*(97)*
Crop inputs	9207	54	*(22)*	34084	211	*(85)*
Crop gross margin	*5597*	*33*	*(13)*	*4724*	*29*	*(12)*
Livestock outputs	417666	2440	*(987)*	653203	4040	*(1635)*
Livestock inputs	186951	1092	*(442)*	279751	1730	*(700)*
Livestock gross margin	*230715*	*1348*	*(545)*	*373452*	*2310*	*(935)*
Whole farm gross margin	*236312*	*1380*	*(559)*	*378176*	*2339*	*(946)*
Other outputs	18835	110	*(45)*	28601	177	*(72)*
Agri-environmental support	11224	66	*(27)*	6129	38	*(15)*
Basic payment scheme	32930	192	*(78)*	30237	187	*(76)*
Total GM incl. subsidies	*299301*	*1748*	*(708)*	*443143*	*2741*	*(1109)*
Labour	52075	304	*(123)*	68714	425	*(172)*
Contract	24624	144	*(58)*	35418	219	*(89)*
Machinery	53284	311	*(126)*	76610	474	*(192)*
General	83921	490	*(198)*	97715	604	*(245)*
Paid rent	25905	151	*(61)*	20422	126	*(51)*
Fixed costs	*239809*	*1401*	*(567)*	*298879*	*1848*	*(748)*

Source: Farm Business Survey data from Rural Business Research

Continued profitability depends on the efficiency of utilising clover-based pastures, in terms of both reduced expenses for fertiliser and improved nutritional quality of forage. The size of organic dairy holdings has continued to increase as a response to the economic pressures affecting conventional and organic producers alike.

Beef/sheep systems

Lowland beef/sheep systems are often similar to mixed dairy/arable systems.

Beef and sheep system gross margin models

The rotation illustrated here is a variant of that used for the mainly dairy example, giving more emphasis to forage production:

7 years grass/white clover ley
1 year winter wheat
1 year winter oats
2 years red clover ley
1 year spring wheat/barley/roots

This rotation and the gross margin model illustration have more in common with the mixed farm survey group shown above.

Upland and hill livestock systems are constrained by climatic and soil quality factors, so that arable cropping is often impossible. These systems will often have no rotation at all, relying predominantly on permanent pasture and rough grazing. A proportion of in-bye land may be suitable for more intensive use with improved grass swards. In this example, forage roots and brassicas are used to cash in on built-up fertility before the new long-term ley is established. The integration of cattle and sheep is critical to parasite control and grassland management in these systems.

Lowland livestock gross margin model — *2023 estimates*

	Area/numbers			*Gross margin*		
Arable crops	ha	*(acres)*		£/ha	*(£/ac)*	£/farm
Winter wheat	15	*(37)*		1007	*(408)*	15105
Spring wheat	5	*(12)*		680	*(275)*	3398
Winter oats	15	*(37)*		882	*(357)*	13223
Spring barley	5	*(12)*		592	*(240)*	2960
Potatoes, maincrop	3	*(7)*		3590	*(1453)*	10769
Carrots	2	*(5)*		3227	*(1306)*	6453
Total arable crops	45	*(111)*		1154	*(467)*	51908
Forage crops	ha	*(acres)*		£/ha	*(£/ac)*	£/farm
Short term leys	30	*(74)*		-181	*-(73)*	-5415
Medium/long term leys	90	*(222)*		-106	*-(43)*	-9495
Permanent grassland	10	*(25)*		-90	*-(36)*	-895
Total forage crops	130	*(321)*		-122	*-(49)*	-15805
Livestock	head	LU		£/hd	£/LU	£/farm
Beef finishing, 18m	64	26		564	1375	36088
Single suckler cows	67	82		504	413	33769
Lowland sheep	500	84		87	517	43439
Total livestock		192			590	113296
Per forage ha		1,48				872
Farm summary	ha	*(acres)*		£/farm ha	*(£/ac)*	£/farm
Arable crops	45	*(111)*	26%	297	*(120)*	51908
Forage/livestock	130	*(321)*	74%	557	*(225)*	97491
Whole farm 2023	175	*(432)*		854	*(345)*	149399
Whole farm 2017	175	*(432)*		737	*(298)*	129027
Whole farm 2014	175	*(432)*		705	*(267)*	123297
Organic Maintenance England				126	*(51)*	21980
Sustainable Farming Incentive Option England est.				200	*(81)*	35000

Upland (LFA-DA) livest. gross margin model — *2023 estimates*

	Area/numbers		*Gross margin*		
Forage crops	ha	*(acres)*	£/ha	*(£/ac)*	£/farm
Forage swedes	5	*(12)*	-299	*-(121)*	-1496
Forage rape	5	*(12)*	-111	*-(45)*	-555
Medium/long term leys	70	*(173)*	-106	*-(43)*	-7385
Permanent grassland	250	*(618)*	-90	*-(36)*	-22375
Total forage crops	330	*(815)*	-96	*-(39)*	-31811
Livestock	head	LU	£/hd	£/LU	£/farm
Beef finishing, 24m	66	49	697	949	46012
Single suckler cows	69	84	504	413	34777
Upland sheep	1000	136	50	371	50480
Total livestock		269		489	131269
Per forage ha		0,81			398
Farm summary	ha	*(acres)*	£/farm ha	*(£/ac)*	£/farm
Whole farm 2023	330	*(815)*	301	*(122)*	99458
Whole farm 2017	330	*(815)*	253	*(103)*	83631
Whole farm 2014	330	*(815)*	256	*(67)*	84347
Organic Maintenance England			59	*(24)*	19560
Sustainable Farming Incentive Option England est.			200	*(81)*	66000

The lowland cattle and sheep Farm Business Survey results for 2021/22 show similar stocking rates and slightly lower livestock gross margins compared to conventional farms. Higher agri-environmental (including organic) support payments and lower fixed costs of all types compensated to yield similar FBI/ha. However, the Basic payments made a critical contribution to overall farm profitability, accounting for more than half of the Farm Business Income (FBI) generated on both organic and conventional samples.

Farm business survey results for lowland cattle and sheep in England, 2021/22

	Organic			*Conventional*		
Farm Nos./Size (ha/acres)	*33*	*87,0*	*(215)*	*247*	*96,4*	*(238)*
Stocking rate (LU/farm ha)	*75,6*	*0,87*	*(0,35)*	*84,8*	*0,88*	*(0,36)*
Financial performance	*£/farm*	*£/farm ha*	*(£/ac)*	*£/farm*	*£/farm ha*	*(£/ac)*
Crop outputs	5029	58	*(23)*	14583	151	*(61)*
Crop inputs	2369	27	*(11)*	8113	84	*(34)*
Crop gross margin	*2660*	*31*	*(12)*	*6470*	*67*	*(27)*
Livestock outputs	51683	594	*(240)*	77723	806	*(326)*
Livestock inputs	12744	146	*(59)*	30408	315	*(128)*
Livestock gross margin	*38939*	*448*	*(181)*	*47315*	*491*	*(199)*
Whole farm gross margin	*41599*	*478*	*(194)*	*53785*	*558*	*(226)*
Other outputs	15748	181	*(73)*	24523	254	*(103)*
Agri-environmental support	9760	112	*(45)*	5935	62	*(25)*
Basic payment scheme	17672	203	*(82)*	17761	184	*(75)*
Total GM incl. subsidies	*84779*	*974*	*(394)*	*102004*	*1058*	*(428)*
Labour	3824	44	*(18)*	5527	57	*(23)*
Contract	5666	65	*(26)*	6846	71	*(29)*
Machinery	15899	183	*(74)*	20269	210	*(85)*
General	26207	301	*(122)*	28353	294	*(119)*
Paid rent	3472	40	*(16)*	7266	75	*(31)*
Fixed costs	*55068*	*633*	*(256)*	*68261*	*708*	*(287)*

Source: Farm Business Survey data from Rural Business Research

The Less Favoured Area (LFA) cattle and sheep farms had higher stocking rates than the conventional farms in 2021/22. Despite this, livestock financial output was lower on the organic farms, but lower variable costs resulted in similar gross margins. Higher agri-environmental (including organic) support payments were offset by higher fixed costs to result in a similar Farm Business Income result per ha.

FBS results for LFA cattle and sheep in England, 2021/22

	Organic			Conventional		
Farm Nos./Size (ha/acres)	*19*	*168,3*	*(416)*	*184*	*150,0*	*(371)*
Stocking rate (LU/farm ha)	*108,7*	*0,65*	*(0,26)*	*88,8*	*0,59*	*(0,24)*
Financial performance	*£/farm*	*£/farm ha*	*(£/ac)*	*£/farm*	*£/farm ha*	*(£/ac)*
Crop outputs	3244	19	*(8)*	5152	34	*(14)*
Crop inputs	3270	19	*(8)*	6604	44	*(18)*
Crop gross margin	*-26*	*0*	*(0)*	*-1452*	*-10*	*-(4)*
Livestock outputs	83667	497	*(201)*	87142	581	*(235)*
Livestock inputs	26400	157	*(63)*	35432	236	*(96)*
Livestock gross margin	*57267*	*340*	*(138)*	*51710*	*345*	*(140)*
Whole farm gross margin	*57241*	*340*	*(138)*	*50258*	*335*	*(136)*
Other outputs	14537	86	*(35)*	11391	76	*(31)*
Agri-environmental support	28522	169	*(69)*	14953	100	*(40)*
Basic payment scheme	40108	238	*(96)*	29652	198	*(80)*
Total GM incl. subsidies	*140408*	*834*	*(338)*	*106254*	*708*	*(287)*
Labour	14083	84	*(34)*	7638	51	*(21)*
Contract	8349	50	*(20)*	4607	31	*(12)*
Machinery	26132	155	*(63)*	18979	127	*(51)*
General	38775	230	*(93)*	24615	164	*(66)*
Paid rent	6540	39	*(16)*	8453	56	*(23)*
Fixed costs	*93879*	*558*	*(226)*	*64292*	*429*	*(173)*

Source: Farm Business Survey data from Rural Business Research

Specialist horticultural systems

Specialist horticultural systems can range from relatively simple, highly-mechanised, field-scale systems supplying pre-packers and wholesalers to very diverse, labour-intensive market gardens providing a wide range of crops marketed directly to consumers and retailers.

Many specialist growers rely on direct marketing, such as farm-gate sales and box schemes, which makes individual crop costings less meaningful and often impossible. In Section 8, we have included a section on yields and estimates of costs and returns for small-scale systems.

Here, the whole farm gross margins of three rotations are illustrated to show in particular the impact of fertility building on the overall returns. The average gross margins from the whole rotation, in these examples, are much lower than the gross margins from the individual cash crops. However, to achieve the various benefits of a good rotation, such as: fertility building, reducing nitrogen

losses, pests and weed control, the lower gross margins from fertility building years or cereals need to be considered as a necessary input cost.

Of the three examples below, the intensive rotation with lettuce shows what is possible with double cropping of high value salad crops, but this is a very specialist operation dependant on a high level of expertise and a secure market for the volume of crops planned and grown on good soils.

Examples of gross margins for some typical horticultural rotations, 2023 estimates

	5-year Rotation			4-year Rotation			4-year Rotation		
Year		£/ha	*(£/ac)*		£/ha	*(£/ac)*		£/ha	*(£/ac)*
1	Grass/Clover	-£181	*-(75)*	Grass/Clover	-£286	*-(119)*	Grass/Clover	-£181	*-(75)*
2	Grass/Clover	-£181	*-(75)*	Spring cabbage	£3.062	*(1276)*	Grass/Clover	-£181	*-(75)*
3	Potatoes	£3.590	*(1496)*	Maincrop onions	£3.684	*(1535)*	Double crop salads	£20.830	*(8679)*
4	Carrots	£3.227	*(1344)*	Sp. Barley	£592	*(247)*	Beetroot	£2.204	*(918)*
5	Sp. Barley	£592	*(247)*						
	Av.	£1.410	*(587)*		£1.763	*(735)*		£5.668	*(2362)*
	Basic payment	£156	*(65)*	Basic payment	£156	*(65)*	Basic payment	£156	*(65)*
	Organic Maintenance	£267	*(111)*	Organic Maintenance	£302	*(126)*	Organic Maintenance	£302	*(126)*

Organic maintenance support based on England rates, using a mixture of payments for rotational land and horticulture, which is limited to 20 ha.

Economic pressures arising from restricted land availability on many specialist horticultural holdings make it difficult to justify putting land down to clover/ grass leys and keeping livestock. While it is possible to replace the fertility building contribution of leys with short and long term green manures such as clover, vetch, mustard and forage rye, often significant quantities of manures, composts and other fertilisers may be brought onto the holding. The source, quantity used and the treatment of these manures may be tightly controlled under organic production standards, but they can be particularly important on smaller, very intensive horticultural holdings. Costs of manure and compost handling and treatment are considered as part of the costs of horticultural production; a typical charge for this has been included in gross margins in Section 8. These costs are not necessarily attributable to the variable costs of an individual crop, since manures and composts release nutrients slowly over several seasons.

FBS results for horticultural holdings in England, 2021/22*

	Organic			Conventional		
Farm Nos./Size (ha/acres)	*9*	*25.4*	*(63)*	*121*	*31.2*	*(77)*
Financial performance	*£/farm*	*£/farm ha*	*(£/ac)*	*£/farm*	*£/farm ha*	*(£/ac)*
Crop outputs	93486	3681	*(1489)*	403141	12921	*(5229)*
Crop inputs	15583	614	*(248)*	157434	5046	*(2042)*
Crop gross margin	*77903*	*3067*	*(1241)*	*245707*	*7875*	*(3187)*
Livestock outputs	0	0	*(0)*	1105	35	*(14)*
Livestock inputs	0	0	*(0)*	579	19	*(8)*
Livestock gross margin	*0*	*0*	*(0)*	*526*	17	*(7)*
Whole farm gross margin	*77903*	*3067*	*(1241)*	*246233*	*7892*	*(3194)*
Other outputs	19936	785	*(318)*	54315	1741	*(705)*
Agri-environmental support	743	29	*(12)*	1149	37	*(15)*
Basic payment scheme	1903	75	*(30)*	4358	140	*(57)*
Total GM incl. subsidies	*100485*	*3956*	*(1601)*	*306055*	*9809*	*(3970)*
Labour	34262	1349	*(546)*	130960	4197	*(1699)*
Contract	3860	152	*(62)*	9652	309	*(125)*
Machinery	11116	438	*(177)*	28152	902	*(365)*
General	20998	827	*(335)*	68731	2203	*(892)*
Paid rent	3859	152	*(61)*	5442	174	*(71)*
Fixed costs	*74095*	*2917*	*(1181)*	*242937*	*7786*	*(3151)*

**A part of this table was derived from data with less than 15 observations in the sample which could reduce the robustness of the results*
Source: Farm Business Survey data from Rural Business Research

Over time, larger organic horticultural holdings have featured in the Farm Business Survey results, but the small sample size still means the results should be treated with caution. Also, the organic group has a lower farm size and a lower level of cropping. The horticultural businesses in the full organic sample achieved an FBI/ha of over half that of the conventional group. Lower output could not be compensated by lower variable costs resulting in lower crop gross margins on the organic farms. Fixed costs including labour were substantially lower on the organic farms, but this did not improve the overall situation.

Organic farming support schemes

In recognition of the contribution which organic farming can make to current agricultural and environmental policy objectives, schemes to support maintenance as well as conversion to organic farming have operated in all of the UK's Devolved Administrations in recent years. Because of the UK's exit from the EU, the overall farm support available structures for England, Wales, Scotland and Northen Ireland are undergoing a transition away from direct payments such as the Basic Payment Scheme (BPS), towards payments oriented towards the provision of public goods, which may include some management practices commonly used by organic farmers. Together this has resulted in considerable uncertainty over the status of organic agriculture in future farm support schemes.

Elements of organic support schemes that deal with conversion to organic farming are described in detail in Section 4. The following paragraphs provide details of the maintenance element of the organic farming grant schemes in the UK regions and the Republic of Ireland (see Section 15 for contact details of agencies). In many cases, both conversion and maintenance schemes can be combined with other agri-environmental schemes, including the additional organic-specific options under Countryside Stewardship in England (for details see Section 13).

England: Countryside Stewardship

The main support for organic farming in England is provided as part of the Mid-Tier Countryside Stewardship (CS). Most elements of organic funding through Countryside Stewardship have been retained since the last edition and the scheme now makes up one strand of Environmental Land Management (ELMS), alongside the Sustainable Farming Incentive (SFI) intended to replace the BPS scheme, and Landscape Recovery (see Section 13 for further details).

The Countryside Stewardship organic land management options are payable on land that is registered as fully organic by a licenced organic certification body (see Section 3) and remain so throughout the duration of the agreement. In most cases the organic land management options can be combined with other CS options on the same parcel of land, in particular the specific organic options focused on wild birds and multi species leys. Payment rates for 2023 agreements are outlined below.

English organic land management grants payment rates 2023 (£/ha)

Option	£/ha	Notes
Improved permanent grassland (OT1)	20	Improved and semi-improved permanent grassland and arable land where the land is to be reverted to permanent improved as part of a CS agreement. At least 10% cover ryegrass and white clover; sward not species rich; ≤ 30% cover wildflower and sedges.* Whole parcel
Unimproved permanent grassland (OT2)	36	Unimproved permanent grassland and rough grazing below the moorland line. Less than 10% cover ryegrass and white clover; sward is species rich; > 30% cover wildflower and sedges.* Whole parcel
Rotational land (OT3)	132	Rotational land parcels that have been cultivated in the preceding 7 years. Whole or part parcel
Horticulture (OT4)	471	Rotational land parcels that have been cultivated in the preceding 7 years; at least 40% horticultural crops and at least 1 crop per parcel. Max 20 ha. Whole or part parcel
Top fruit (OT5)	1920	Top fruit and permanent bush crops. Orchards used in production of alcoholic drinks or not in commercial production are not eligible. Whole parcel
Enclosed rough grazing (OT6)	69	Land registered as 'in conversion' or 'fully organic'. Rough grazing parcels < 15ha, within SDA, above moorland line. Cannot be used on land subject to other land management or conversion options.

** At least two criteria should apply*

As with organic conversion grants, at the time of writing, payments to grant holders have finally been assured for 2024 agreements. For full details on criteria, applications and future funding rounds see:

www.gov.uk/guidance/organic-farming-how-to-get-certification-and-apply-for-funding.

Technical advice for all CS schemes is provided by the Natural England Enquiries Team via:

- Email: enquiries@naturalengland.org.uk
- Telephone: 0300 060 3900

In addition, there are number of options under the newly launched Sustainable Farming Incentive (SFI). SFI is intended to more than replace the BPS with a

range of payment options, intended to enhance the public good delivery of farming. Under SFI Farmers will get paid for taking actions that support food production and improve farm productivity and resilience, while also protecting and improving the environment. The 23 actions on offer cover existing themes including soil health and moorland, as well as new actions on hedgerows, integrated pest management, nutrient management, farmland wildlife, buffer strips, and low input grassland (see also Section 13). SFI 2022 has now closed and anyone who has previously applied for SFI soils payments will have to reapply. The RPA has sent letters out to eligible BPS farmers inviting them to express an interest in SFI 2023. SFI options run for three years rather than 5-year agreements under CS. For further details on the scheme see:

www.gov.uk/government/publications/sfi-handbook-for-the-sfi-2023-offer

Many SFI options can be combined (stacked) with organic CS options, unless they are supporting the same actions with SFI payments intended to be less prescriptive than organic. This provides additional financial support for organic farmers. All farms including those with CS and ES agreements can claim the SAM1, HRW1, IPM1, NUM1 options. The Farm Consultancy Group has prepared a technical leaflet for OF&G on how the two schemes can be combined, illustrated with a couple of case studies[3]. This lists Soils Action (SAM 1 to 3), Nutrient Management options (NUM1, 2 and 3) and some IPM options (IPM 1, 3 and 4) as relevant for the organic farmers.

Scotland: Organic Farming – Maintenance

The Scottish Government offers maintenance and conversion support for organic farms throughout Scotland within the Agri-Environment Climate Scheme (AECS), with the aim of expanding organic production in Scotland for environmental benefits. To be eligible land should be registered as fully organic with a certification body (see Section 3) and must remain so throughout the duration of the contract, which is for up to a maximum of five years.

Payments for organic conversion under AECS are no longer capped at 1000 hectare and after the first two years of conversion, fully converted grant holders can expect payments of £65/ha for arable land, £200/ha for vegetable or fruit production, £55/ha for improved grassland, and £8.50/ha for rough grazing.

---

3. OF&G Technical Leaflet TL130 Sustainable Farming Incentive and Countryside Stewardship for Organic Farmers 15 September 2023
https://ofgorganic.org/docs/sfi-and-cs-for-organic-farmers-sept-2023.pdf

Wales: Glastir Organic

Glastir Organic is the organic support element under the Welsh Government's Glastir Scheme, which is due to come to an end in 2023. From 2025, the organic farmers will be able to apply for the new Sustainable Farming Scheme, but this leaves a gap year for support in 2024. The Sustainable Farming Scheme is expected to have a strong focus on delivering more sustainable food production, climate change mitigation and adaptation, ecosystem resilience, as well as promoting the cultural value of the Welsh countryside and it is expected that organic farmers should be able to qualify easily. As a result, there are currently no plans for a replacement organic maintenance scheme.

Glastir Organic support has not been open to new applicants in recent years. The support covered both conversion (see Section 4) and maintenance with each approved parcel receiving a 5-year schedule of payments dependent upon the conversion or maintenance status of the parcel upon entry to the scheme, and according to three area-based payment rates. Payments were capped to a maximum of 400ha total land area and 20ha at payment rate 1, with the minimum area of land eligible set at 3ha. The scheme opened again for applicants for organic conversion in 2022 and there has been budget allocated for payment of awarded grants in 2023.

Northern Ireland

Similarly to Wales, Northern Ireland is transitioning towards a new funding structure for agriculture and many existing funding opportunities will not be available in 2024 as the new scheme is developed and implemented. This includes the Environmental Farming Scheme Wider Stand Alone Options that previously included management payments for a range of organic options. At the time of writing it has not been confirmed whether a specific organic management option will be available.

Existing grant holders from 2020/21 would have received the following payment rates for organic management:

Payment rates for fully converted farms in Northern Ireland in 2022/21 (Years 1 - 5)

	Years 1-5
Organic arable	£53 per ha up to 60ha then reduced to £20/ha
Organic grassland	£53 per ha up to 60ha then reduced to £20/ha
Organic horticulture	£197 per ha up to 6ha then reduced to £53/ha
Organic horticulture top fruit	£197 per ha up to 6ha then reduced to £53/ha

** At least two criteria should apply*

Republic of Ireland: Organic Farming Scheme[4]

Ireland's Organic Farming Scheme (OFS) provides support on a 5-year basis for organic conversion as part of the new Agri-Climate Rural Environment Scheme (ACRES). The scheme accepted applications for the 2023 period. Grant holders will receive larger payments for a conversion 2-year conversion period as well as payments for management of a fully converted organic system for the subsequent 3 years. An additional payment of €1,400 for organic license holders to cover administration and training costs etc. will also be available in this period. The payment rates per hectare are as follows:

Irish payment rates 2023 for fully converted farms

	Rates for 1-70ha (€/ha)	Rate for >70ha) (€/ha)	Notes
Horticulture	600	30	Organic horticulture producers, with an organic horticulture area of one hectare or more, are eligible provided that at least 50% of the area eligible for organic payment is cropped each year.
Tillage	270	30	Organic tillage producers, with an organic tillage area of six hectares or more
Dairy	300	30	Dairy producers, with an organic area of six hectares or more.
Dry stock (beef & sheep) and all other holdings	250	30	Applicants with 3 hectares or more of utilisable organic agricultural area are eligible.

Direct payments

Basic Payment Scheme (BPS)

Historically, the BPS has provided annual payments to farmers for eligible agricultural activities. It was considered a safety net for farmers and landowners, providing income support in an increasingly globalised and highly volatile global food and drink market. Following the UK-exit from the EU, the UK's administrations have chosen different pathways when it comes to phasing out and/or reforming direct payments under the BPS.

4. www.irishorganicassociation.ie/farming-2/organic-farming-scheme-and-grant-aid-scheme/

In **England** land is split into three regions – Lowland, Severely Disadvantaged Areas (SDAs) and Moorland – with different payment 'entitlements'.

BPS scheme entitlements for England, 2023[5]

Payment region	£/ha (≤128.64ha)	£/ha (128.64-214.39ha)	£/ha (214.39-643.18ha)	£/ha (over 643.18ha)
Lowland	151.59	139.93	116.6	104.94
SDA	150.52	138.94	115.79	104.21
Moorland	41.57	38.37	31.98	28.78

The direct payments currently paid through the BPS are being steadily phased out through to 2027 as the government introduces the Environmental Land Management schemes. As part of this process, BPS payments will be replaced with delinked payments based on a farmer's average BPS payments for the 2020 to 2022 scheme years. These payments will be reduced each year as they are phased out. Farmers must be eligible for BPS payments and must have already claimed for BPS in the 2023 scheme year to receive delinked payments for 2024 to 2027. Existing claimants can calculate their expected reduction in BPS payments via this link: https://calculate-direct-payment-reductions.defra.gov.uk/

The Sustainable Farming Incentive (SFI) is expected to fully replace BPS as a means of income support beyond 2027. The fully range of SFI options are expected to be available by 2025. Further details on how existing and available SFI options could compliment CS organic conversion and management payments is provided in Section 13.

In **Scotland** it has been confirmed that the existing framework of support in will continue in 2023, 2024 and mostly for 2025 but with new conditionality to be introduced to the Basic Payment Scheme (BPS) from 2025. The exact details of conditions to be met for Tier 1 Basic Payments under the forthcoming program were not available at the time of print but they are expected to build on existing greening requirements which already has conditions to be met for permanent grassland, crop diversification, and establishing Ecological Focus Areas (EFAs). The Scottish Government has published a route map for agricultural payments:

www.ruralpayments.org/topics/agricultural-reform-programme/arp-route-map/

5. https://townsendcharteredsurveyors.co.uk/farm-quota/entitlements/uk-bps-entitlements-user-guide/

Scotland has established three payment regions: Arable and Grassland (Region 1), Rough Grazing (Region 2) and Poor Rough Grazing (Region 3). BPS payments are based upon the value of entitlements for each region set each year.

BPS scheme flat payment rates (£/ha) for Scotland 2023

Payment region	BPS	Greening	Combined
1 Arable/grass	£147.64	£75.92	£223.56
2 Rough grazing	£32.51	£12.85	£45.36
3 Poor (LFA A) RG	£9.46	£4.30	£13.76

The extension of BPS has been confirmed in **Wales** for 2023 and 2024 with existing Greening Requirements forming 30% of total payment. Only Wales has an option for redistributive payments, which are applied to the first 54ha of any claim, designed to support smaller farmers. Rates have remained the same as in 2021 and 2022, with the total budget for direct payments totalling £238 million. Within this budget farmers have received approximately £175/ha with a re-distributive payment on the first 54 ha of £112. Further information regarding applications can be viewed here: https://tinyurl.com/BPS-Wales

Northern Ireland also implements a flat-rate payment, which is expected to be implemented as normal until 2025. In 2025, a direct payment will still be available for farmers but with new conditions being introduced throughout the year in preparation for Northern Ireland's new Farm Sustainability Payment scheme launching in 2026. Further information on the timeline for changes to Northern Irish farm funding can be viewed here: https://tinyurl.com/DAERA-BPS

In the **Republic of Ireland**, a new CAP regime has been implemented from 2023. We have not been able to include details in this edition.

Cross compliance regulations state that eligibility for direct payments (BPS) depends on the land being farmed according to the Statutory Management Requirements (SMR) and being kept in Good Agricultural and Environmental Condition (GAEC). Farms receiving the direct payments must be open to inspection by the relevant authorities. Organic producers are generally well able to meet these rules. Since 2020, farmers in Northern Ireland and Scotland are no longer required to follow cross compliance regulations.

Greening component of the Basic Payment Scheme

IIn order to qualify for the last 30% of the Basic Payment Scheme (BPS), farmers had to be Greening-compliant as well as Cross-compliant. From 2021, greening

requirements for farmers in England were removed. Farmers in Scotland and Northern Ireland no longer need to meet the crop diversification rules. The new schemes under development will replace this in future.

The Greening requirements consisted of three main elements:

Crop diversification: On arable areas of 10-30 ha, at least two different crops must be grown, with no crop more than 75% of the arable area. If more than 30ha are grown, at least three crops must be grown, with not more than 75% of the area covered by the main crop, or 95% covered by two crops. Spring and winter plantings of the same crop are considered to be different crops, and temporary grass/clover leys and other legumes/green manures and fallows also count as crop types for this purpose. The crop diversification requirement does not apply to farms with less than 10 ha arable land or more than 75% temporary grass or fallow within the arable area, or more than 75% permanent and temporary grassland on a whole farm basis, provided that the remaining cropped area does not exceed 30 ha.

Permanent pasture (including leys > 5 years old): Already a requirement under cross-compliance at national level, the area of permanent grassland should not fall more than 5% below a set reference level. England, Wales and Scotland implemented this at the national level. Cross-compliance, EIA and organic standard restrictions on ploughing up environmentally-sensitive pastures still apply.

Ecological focus areas: 5% of the arable area (i.e. non-permanent grassland area) should be managed primarily for non-agricultural purposes, including landscape features (hedges and trees), buffer strips, fallow land, protein-fixing crops (legumes), short-rotation coppice and agroforestry, with the possibility of applying different weightings to different components. Each region of the UK has set its own criteria to define EFAs.

Organic farmers qualified automatically for Greening, although for farms which are only part-organic, the normal Greening compliance requirements still need to be met on the non-organic land. Farms with more than 75% permanent and temporary grassland are also exempted from most Greening requirements subject to some constraints (e.g. not more than 30ha cropped land).

Payments for Less Favoured Areas (LFAs)

Scotland is currently the only region in the UK to offer an LFA support scheme (LFASS), which is set to continue into 2024. Under the current LFASS, payment rates are set according to four grazing categories which are grouped into 'more disadvantaged land' (categories A and B) and 'less disadvantaged land'

(categories C and D). These two categories are then assigned payment rates within three land areas according to land fragility, with maximum stocking densities applicable to each category. The payment rates and required stocking densities for the scheme are as follows:

Scotland LFASS payment rates (£/adjusted ha), 2023

	Grazing category	Standard area	Fragile mainland area	Very fragile island areas	Stocking densities (lu/ha)
More disadvantaged land	Category A	52.15	62.10	71.35	Up to 0.19
	Category B	52.16	62.10	71.35	0.2 to 0.39
Less disadvantaged land	Category C	34.12	54.51	63.00	0.4 to 0.59
	Category D	34.12	54.51	63.00	0.6 or more

For full guidance on Scottish LFA support, visit: https://tinyurl.com/LFASS-pay

The LFASS is under review and will be reformed in 2025 to enable wider participation of active farmers and crofters who have previously been ineligible for the scheme.

Further reading

More in-depth overviews of the economic performance of organic farms in various countries, and the factors affecting their profitability, can be found in the following reviews, which cover key insights and principles that remain relevant today even if the data are outdated.

Lampkin N, Padel S (eds) (1994) The Economics of Organic Farming - an international perspective. CAB International, Wallingford.

Offermann F, Nieberg H (2000) Economic performance of organic farms in Europe. University of Hohenheim, Stuttgart.

Financial performance, benchmarking and management of livestock and mixed organic farming, IOTA PACARes Technical Leaflet (2) www.organicresearchcentre.com/?go=IOTA&page=leaflets

Lampkin N, Gerrard C, Moakes S (2014) Long term trends in the financial performance of organic farms in England and Wales, 2006/07-2011/12. Report commissioned by the Soil Association. Organic Research Centre, Newbury.

Scott C (2023) Farm Business Survey 2021/2022 Organic Farming in England. Rural Business Research at Newcastle University, Newcastle. This reports and those previous years are available at: https://www.ruralbusinessresearch.co.uk/

All years of the Organic Farm Income in England and Wales reports from Aberystwyth University can be accessed via http://tinyurl.com/OFIreports

SHIPTON
MILL
EXCEPTIONAL MILLERS · OF QUALITY FLOUR ·
Welcome to Shipton Mill, the home of regenerative, organic flour
We started Shipton Mill in 1979, to mill flours from the most fantastic grains, working from day one with like-minded farmers who promote biodiversity and who value and seek to look after the soil.
Our flours are for all bakers and cooks, united by their passion for good foods

Soil management, soil analysis and rotation design

This section summarises the key principles that govern organic crop production in practice. The principles and practice of organic crop husbandry are described in detail in the IFOAM organic principles (Section 1), the European Organic Regulations, Defra and standards documents of the Control Bodies (Section 3). See also further reading at the end of this Section.

Soil fertility and soil management[1]

Key principles of organic crop production include maintaining and enhancing soil fertility and soil life, nourishing the plants through the soil ecosystem, preventing and combatting soil erosion and compaction and minimising the use of external inputs. In organic farming, soil fertility should be seen mainly as a result of biological processes and recycling nutrients around the farming system, not the application of synthetic nutrients. A fertile soil reacts actively with the plants; it structures itself and is capable of regeneration. Physical qualities can be recognised through observation, for example using a spade test. Biological qualities can be seen through their transforming/recycling activity, and the occurrence and visible evidence of life forms in the soil. Chemical qualities can be determined through measuring individual macro- and micronutrients and the pH value (see *Soil and plant analysis* below).

Soil cultivation should support soil organisms in building a soil structure that allows for deep rooting by crops and provides adequate aeration and drainage so that plant roots can exploit available nutrients in the full soil profile. This is particularly important, given that the application of highly soluble nutrient sources to the soil surface is avoided. Cultivations should aim to maintain the biologically active surface layers in the top 15-20cm (6-8"). Shallow ploughing or non-inversion tillage is normal practice, in combination with sub-soiling if necessary. But it must be remembered that the chosen method of cultivation will also need to fulfil other objectives, in particular weed control. Some standard labour requirements and machinery costs for different soil cultivations are given in Section 12.

---

1. The Basics of Soil Fertility: Shaping our relationship to the soil.
https://tinyurl.com/ORC-soilfertility
Soil analysis and management IOTA Technical leaflet 4: https://tinyurl.com/IOTA-soilanalysis

Soil management should also include the regular return of carbon-containing organic matter, derived from crop residues, straw incorporation, green manures, livestock manures or compost. Organic matter plays a key role in maintaining the biological activity of the soil, providing soil microorganisms and earthworms with energy to make nutrients available to crop plants, and in maintaining a stable soil structure, and for carbon sequestration. Organic farmers should aim to maintain or increase soil organic matter (SOM) in the long term.

The need to increase levels and the potential for carbon sequestration in a given situation depends on current levels and the soil type. The aim should be to increase SOM levels in soils with low organic matter e.g., less than 3%. However, this may be relatively difficult in sandy soils and it should be noted that soil carbon levels may go down as well as up, depending on the stage of rotation and changes in management.

Soil fertility and crop productivity may be more dependent on the input of high carbon materials to the soil for the effect that they have on stimulating soil biological activity rather than the SOM level *per se*.

In order to monitor changes in SOM over time it is essential that rigorous, standardised monitoring procedures are in place, including sampling methods, sampling location, to a depth of 50 cm and annual monitoring using identical analytical methods, including bulk density, over many years. Increases in the order of 1 – 3 % of the existing total SOM, per year may be possible in depleted situations.

There are currently no government backed, regulated soil carbon sequestration markets.

Soil and plant analysis[2]

Soil analysis is an essential tool for organic farmers; it provides important information in order to ensure that soil nutrient levels are satisfactory for optimum crop and forage production and to avoid nutrient deficiencies in crops or livestock. Regular visual observation of soils and plants is also important, for example through spade assessment. If used in conjunction with information on soil type, pH and organic matter and with knowledge of the planned

---

2. GREATSoils Soil Health Scorecard: https://ahdb.org.uk/greatsoils
   Soil analysis and management, IOTA Technical leaflet 4. https://tinyurl.com/IOTA-soilanalysis
   FiBL Factsheet No 1349 Soil and climate
   https://www.fibl.org/fileadmin/documents/shop/1349-soil-and-climate.pdf

cropping and nutrient budgeting, the soil analysis will contribute to providing recommendations on the use of permitted fertilisers and the targeted use and application rates of manure to particular fields and rotation design.

Soil sampling is normally undertaken annually in more intensive horticultural systems, at least once every rotation in arable fields, and every four to five years in grassland. In addition, sampling the same one or two fields every year provides an invaluable assessment of long-term trends in soil organic matter and nutrients, indicating if the farming system is sustainable in the long term.

There are a number of different analytical services available for organic farming. The standard analysis for pH, P, K and Mg is the most commonly used; the results are expressed as Index values, and this generally meets the needs of routine analysis. It is important to choose an analytical technique that is suited to the soil type (standard methods differ in England and Scotland because soil type is generally different). In addition, there is the option of soil organic matter analysis which is useful for long term monitoring. If the aim is to do comparison over time, the same analytical technique and preferably the same lab should always be used. Soil type information and trace element analysis can also be important, the latter to assess the risk of possible crop or livestock deficiencies. Standard analysis costs (excluding trace elements) are in the range of £15 - 18 per sample (£25 - 31 including organic matter). Most laboratories in the UK do not give interpretation or recommendations relevant to organic production.

There are also several other analytical services offered by specialist laboratories including the Soil Health Test by NRM laboratories £77/sample covering P, K, Mg, pH, organic matter, texture and respiration rates (the latter may give indications on active soil life); Soil Mineral N £15/sample; Base Cation Saturation Ratio (Albrecht), £85/sample; Soil Food-Web analysis £195/sample, and other soil biology assessments ranging between £50 for active bacteria and fungi up to £200/sample for more complex analysis. Compost tea analysis is also available for £75. There is currently little research evidence to support the routine use of these more complex, biological analytical services.

Plant tissue analysis for N, S, P, K, Ca, Mg, Mn, Fe, Cu, Zn and Bo is £64.50 per sample. Tissue analysis is useful to monitor crop levels and diagnose yield problems or deficiency symptoms.

Rotation design

The rotation is the core of most organic cropping farms, based on the principle that diversity and complexity provide stability in the agricultural ecosystems. Beneficial interactions between enterprises and between the farm and the

external environment should be exploited fully. The role of the rotation is therefore to:

- ensure sufficient crop nutrients and minimise their loss;
- provide a self-sustaining supply of nitrogen through the use of legumes;
- minimise and help control weed, pest and disease problems;
- maintain soil organic matter and soil structure;
- provide sufficient livestock feed where necessary; and
- maintain a profitable output of cash crops and/or livestock

As conditions vary between farms, there is no 'blueprint' organic rotation, but the tables below provide some guidance. Even if the market changes quicker than a planned 4–5 year rotation, there is a need to respond without endangering fertility building as the basis for successful organic production. There may be no rotation at all in the case of all-grass farms or perennial cropping. Instead, diversity is achieved in space through species and varietal mixtures rather than over time.

Proportions of crop categories in organic rotations

Crop category	Min (%)	Max (%)
Forage, green manure legumes & leys	30 (a)	100
Cereals	0	50 (b)
Other fodder crops	0	33 (c)
Grain legumes & oil seeds	0	25 (b)
Roots & vegetables	0	75 (a, b, c)

Likely consequences of exceeding the limits are:
a. not enough fertility building,
b. problems with specific diseases and the build- up of weeds and
c. reduction of organic matter in the soil because of frequent cultivation.

Intervals between crops in rotation design

Regular return of the farm's own manure will be important on mixed farms. The aim of achieving crop production objectives through management of the soil/ farm ecosystem, rather than reliance on external inputs, places constraints on the choice and combination of crop and livestock enterprises that may have significant implications for the overall performance of the farming system. The rotational and other characteristics of some common organic farm types and their typical whole farm gross margins are illustrated in Section 5.

Crop	Interval (yrs)	Crop	Interval (yrs)
Red clover	5-6*	Fodder & sugar beets	4-5
Lucerne	5	Swedes & turnips	3-4
Oil seed rape	3-4	Brassicas	4
Oats	3-4	Onions & other alliums	4-6
Peas & beans	4-5	Potatoes	5-6

*max. 3 years continuous use, after that period the break of 5 years applies

The objective of reliance on farm or locally derived renewable resources in organic farming means that, where possible, fertility should be obtained from within the farm system. In the case of nitrogen, which is freely available from the atmosphere, biological nitrogen fixation provides the basis for nitrogen self-sufficiency. This requires an appropriate proportion of legumes in the rotation and good management of crop residues and livestock manures to avoid leaching. The information available to enable organic farmers to better estimate the nitrogen supply from fertility-building crops has improved and we have included some below.

Crop nutrition and nutrient budgeting[3]

Organic farmers aim to balance the inputs and outputs of nutrients in the farm or horticultural system. Despite the aim to improve the recycling of all crop nutrients, there will always be some net export from the holding as crops and livestock are sold. The first aim should be to minimise unnecessary losses from the system by avoiding leaching and erosion. Attention should also be paid to the sale of nutrients from the farm, for example, high potassium losses from the system through the selling of straw and conserved forage crops. Secondly, some reliance can be placed on the release of nutrients from soil minerals through normal soil formation processes, good soil structure and growing crops that improve nutrient availability. Purchase of livestock feeds can also add a substantial amount of nutrients to the cycle.

---

3. The role, analysis and management of soil life and organic matter in soil health, crop nutrition and productivity. https://tinyurl.com/IOTA-soil-life
   Managing Manure on Organic Farms. EFRC & ADAS. https://orgprints.org/24819

SHIPTON MILL

Nutrient budgeting[4]

Nutrient budgeting provides a useful means of assessing whether the farm is in balance and provides information for rotation planning and manure and permitted fertiliser input. Budgets can be used both to assess potential deficits or surpluses of nutrients and to provide guidelines for nutrient management decisions. Nutrient budgets are commonly used as a tool in the following circumstances:

- To allow farmers and growers to make better use of available nutrients;
- To design and evaluate the viability and sustainability of arable and horticultural crop rotations;
- To provide information to Control Bodies on the need for restricted fertiliser inputs; and
- To indicate likely surpluses of nitrogen on the farm and therefore the risk of losses by leaching to ground and surface water, especially in Environmentally Sensitive Land Management Sites or Nitrate Vulnerable Zones.

Inputs and outputs for each nutrient and the surplus or deficit for the whole farm are calculated. Although the types and amounts of inputs and outputs of nutrients vary between fields, farming systems and regions, nutrient budgets provide a framework that can be applied systematically across a range of systems and scales, for single fields, across complete rotations or for whole farm systems. Published figures are available for the nutrient content of harvested crops and for the inputs used (see also p 226 for nutrient content of organic manures). Other standard values that can be used for nutrient budgets can be found in the IOTA Technical leaflet *Guide to Nutrient Budgeting on Organic Farms*[5].

Measurements and estimates have also been made of the nitrogen fixed by leguminous crops. Although published figures are based on laboratory analysis of a large number of samples and will be correct on average across the UK, there are considerable variations in crop quality and yield, nutrient contents of manures and in the actual amounts of nitrogen fixed by legumes in any season. Consequently, budgets cannot be used to give exact recommendations and the results should be interpreted carefully. The example of a whole farm nutrient budget in the table below includes the nutrients leaving and entering

---

4. IOTA Technical Leaflets: Compost - the effect on nutrients, soil health and crop quantity and quality. https://tinyurl.com/IOTA-compost
5. *A guide to nutrient budgeting on organic farms* IOTA Technical Leaflet 6 https://orgprints.org/id/eprint/31654/

the farm gate over a 12-month period. This example indicates a small but acceptable surplus of nitrogen, a satisfactory balance of phosphate but a deficit of potassium, which will need to be addressed particularly on sandy soils by operating a less intensive rotation or bringing in high K manure or slurry or purchasing a permitted K fertilizer, such as potassium sulphate.

Example of annual farm gate nutrient budget for 60 ha mixed cattle and sheep

	Input/output data				Total		kg nutrients in/out		
	N	P	K			In/out	N	P	K
Nitrogen fixation	kg/hectare				ha		kg/farm		
White clover ley,	150				10		1500		
White clover ley, cut	150				10		1500		
Permament pasture	150				10		1500		
Spring beans	100				10		1000		
Crop off-take	kg/t			t/ha	ha	t	kg/farm		
Winter wheat grain	-13	-3	-4	5	10	50	-650	-150	-200
Straw sold	-4	-1	-9	2	10	20	-80	-20	-180
Winter oats grain sold	-14	-3	-4	4	10	40	-560	-120	-160
Straw used on farm	0	0	0						
Spring beans	-58	-8	-34	3	10	30	-1740	-240	-1020
Inputs	kg/t			t/ha	ha	t	kg/farm		
Seed wheat purchase	13	3	4	0.2	10	2	26	6	8
Seed oat own	0	0	0	0.2	10	2			
Seed beans purchase	58	8	34	0.25	10	2	116	16	68
Rock Phosphate	0	118	0	0.3	15	4.5		531	
Manure purchase	6	1.5	5.5			50	300	75	275
Purch. feed (rape	58	10	13			5	290	50	65
Straw purchase	4	1	9			0			
Atmospheric deposition	25	0.1	4		60		150	4	24
Livestock	kg/t			d	ha	t	kg/farm		
Cattle store purchase	25	8	2	0.3	40	12	300	96	24
Cattle sold	-25	-8	-2	0.6	40	23	-575	-184	-46
							N	P	K
Total farm nutrients in/out						kg	*3077*	*64*	*-1142*
Av. nutrient balance per hectare						kg/ha	*51*	*1*	*-19*

If nutrient exports through sales exceed the purchases and the expected rate of natural regeneration, then it will be necessary to supplement nutrients like potassium, phosphate, calcium, magnesium and trace elements from external, permitted sources. These should preferably be in an organic form, or in a less soluble mineral form, so that the nutrients are released and made available to plants by the action of soil microorganisms. In this way, luxury uptake by crop plants can be avoided. This may not be possible in all cases, for example for potassium. The production standards allow restricted use of potassium inputs (in various forms) in cases of demonstrable need and matched to soil type and crops planned. A list of the inputs for soil improvement and crop nutrition permitted for use on organic farms is available from the Control Bodies. Some examples with prices are also shown below.

Permitted fertilisers and soil amendments

Fertilisers: Nutrient management in organic systems has a longer perspective than a single crop or season, due to the use of crop rotations and the inclusion of animals within the system. The results of soil analyses and nutrient budgeting should be considered before using permitted fertilisers.

A range of organic fertilisers, including composts, poultry and other livestock manures, green waste composts and biogas digestate are available.

Some manures and products listed on the following page may only be used on a restricted basis and approval should be sought from the relevant certifying body before use. Prices of bulky products will vary considerably depending on the size of the order, whether in bulk or bags and the distance from the depot. Addresses of suppliers of the inputs listed can be obtained from the Control Bodies and advisory organisations[6]. Commercially composted materials are widely available, due to a re-structuring of the waste management sector in the UK. Their use is permitted within agreed limits and if a satisfactory analysis has been obtained prior to their use.

Growing media for potting and seedlings: Potting and seed composts for raising transplants must also be produced using permitted ingredients, for which the Control Bodies can provide addresses of suppliers. All transplants and seedlings raised for organic production must be raised on an organic holding. Organic blocking and module composts are available at £5.00-9.50 per 60-80 litres plus delivery, but there can be considerable problems in obtaining small quantities because of prevailing minimum order requirements. The quantities

6. https://tinyurl.com/SA-approved-inputs
https://ofgorganic.org/useful-info/approved-suppliers?q=

required are higher than for conventional plant raising since larger cell sizes are used. Further research and development work is underway to improve the quality of these products, particularly with the use of peat being phased out, provisionally by 2025 for organic and 2028 for all production.

Standard conventional fertiliser recommended application rates are available in the Defra Fertiliser Manual (RB209)[7]. Some of the data is useful to organic farmers in that it provides nutrient analysis information for manures and rock fertilisers and crop off-take data. It is not relevant for calculating organic crop nutrient inputs, which is based on fertility building, recycling and enhanced biological activity along with the use of slow release mineral fertilisers.

Price guide for fertilisers and soil amendments

Product	Price (excl. VAT) Bulk (£/t)
Nitrogen sources	
High N, granular (13%N 3%P 3%K)	350-600
Phosphate sources	
Soft rock phosphate (e.g. Gafsa 27-33% P_2O_5)	260-520
Potassium sources	
Sulphate of potash, granular (50% K_2O, 45% SO_3)	550-600
Sulphate of potash & magnesium salt (30% K_2O, 10% MgO, 42% SO_3)	550-600
Kainit (11% K_2O, 5% MgO, 27% Na_2O, 10% SO_3)	250
High K, granular potash (24% K_2O, 15% SO_3)	200
Calcium (lime) sources	
Ground chalk/limestone spread (typically 50% CaO)	25-30
Calcium carbonate	450
Magnesium sources	
Epsom salt (16% MgO, 32% SO_3)	20-40/25 kg
Magnesium rock (e.g. Kieserite 25% MgO, 50% SO_3)	300
Compound fertilisers	
N-P-K tailored compounds	420-460
Trace elements	
Mn,Cu,Zn,Fe,Se,B (concentrate powder)	43/kg (0.03/litre-liquid), 1kg for 1200l)
Seaweed foliar (liquid)	5.2/litre
Seaweed meal	1400

7. https://ahdb.org.uk/knowledge-library

SHIPTON MILL

Seeds, seed potatoes and transplants

The retained EU Organic Regulations require that organic producers use organic seed and vegetative reproductive materials grown by registered organic seed producers/plant raisers. It defines organic seed as having been produced from parent plants that have been organically raised for at least one generation, i.e. grown organically from seed to seed for annual and biennial crops. Currently, no more comprehensive standards exist for the production of organic seed. The issue of seed-borne diseases requires careful attention and the use of resistant varieties and development of testing protocols is likely to become more important. Organic seed production requires a professional approach[8]. Home saving of seed can reduce costs, but it is imperative that it is tested for disease and admix. Where organic seeds or plants are not available, derogations may be granted. A European internet database with specific UK sections (www.organicxseeds.co.uk) shows the availability of organic seed. If organic seed is available for different varieties, producers must justify the use of non-organic varieties. Poor quality of organic seed is not usually recognised as a reason for granting derogations. All seeds and vegetative material must also have been produced without the use of any GMO derivatives.

Growing vegetable, herb or flower seeds as a commercial crop has barely featured within UK organic farming in many decades. A renaissance in seed growing has been happening around the world, particularly in the organic sector to better meet the needs of organic farmers. However, the UK still imports approximately 80% of its organic vegetable, herb and flower seed and with demand still at a high since the global Covid-19 pandemic, and challenges due to the UK-exit from the EU, shortages have been known to occur. The Gaia Foundation's Seed Sovereignty Programme of UK and Ireland supports organically produced and open pollinated seeds by providing training, resources and information for small-scale commercial growers to venture into seed production. www.seedsovereignty.info/

Currently, there is sufficient organic seed available for most cereal crops and seed potatoes, i.e., 100% organic seed should be used. Unforeseen growth in the area of individual crops could, however, lead to shortages for some crops. Provided grass, forage and green manure mixtures contain a minimum of 70% organic seed, there is no need to apply in advance, but the derogation must be applied for and approved before the next annual inspection. The situation with vegetable seeds is variable, but many of the varieties required by larger buyers are not available as organic seed, resulting in use of conventional seed under derogation. This includes the production of transplants for these crops.

Crop protection

Weed control[9] is achieved primarily through preventive measures, including rotation design, soil cultivations, crop variety selection, under-sowing, use of transplants, timing of operations and mulching with crop residues or other materials. Direct intervention to correct weed problems should be seen as secondary. Options for this include mechanical, thermal and biological controls of different types. It is important to match the use of appropriate weeding strategies and equipment to the weed type. In practice, the need for significant additional weed control in arable crops and grassland (in addition to cultivation around crop establishment) is relatively low, although in some cases perennial weeds (e.g. some grass weeds, docks and thistles) can become problematic. The use of inter-row hoeing techniques is growing in popularity. They rely on wide row widths and passing the hoe through the growing crop to achieve weed control and may have some additional benefits of mineralising nitrogen for the growing crop (costs for mechanical weed control are illustrated in Section 12). For cereals, pulse, and root and vegetable crops mechanical control with or without vision guidance systems are available, but some hand-weeding may be unavoidable. The number of hours required will depend on the amount of weeds, level of mechanisation of the holding, weather conditions and timeliness of control.

Pest and disease control should also be achieved primarily through preventive measures, including crop hygiene, balanced crop nutrition, rotation design, variety selection for resistance, habitat management to encourage pest predators (see Green Manures in Section 9 for costs) and organic manure/ compost to stimulate antagonists to soil-borne pathogens. Where direct intervention is required, production standards permit a number of biological controls such as *Bacillus thuringiensis* for cabbage white caterpillar control – and certain non-synthetic pesticides and fungicides.

8. Organic Plant Breeding www.eco-pb.org
   Liveseeding - transforming organic seed systems https://liveseeding.eu/
9. http://www.gardenorganic.org.uk/weed-management

SHIPTON MILL

Price guide for pest and disease control inputs

Product	Notes	Price (excl. VAT) £
Insecticides		
Pyrethrum	(Spruzit)	140-170/l
Potassium soap	(containing fatty acids)	12-17 per l (conc)
Biological controls		
Amblyseius cucumeris	(thrips control - indoors only)	*The price of biological control products is highly dependent on the characteristics of the farm itself, the nature of the pest problem and the quantity required. Generally, this will be a bespoke solution.*
Aphidius colemani	(aphid control - indoors only)	
Aphidoletes aphidimyza	(aphid control - indoors only)	
Bacillus thurungiensis	(caterpillar control - in/outdoors)	
Cryptolaemus montrouzieri	(mealybug control - indoors only)	
Dacnusa sibirica	(leaf miner control - indoors only)	
Diglyphus isaea	(leaf miner control - indoors only)	
Encarsia formosa	(whitefly control - - indoors only)	
Hypoapsis	(sciarid larvae control - indoors only)	
Phasmarhabditis	(nematode for slug control- - indoors onl	
Phytoseiulus persimilis	(red spider mite control - indoors only)	
Fungicides		
Vegetable oils	(aduvant-helps active ingredient stick to plants)	3.5/1
Sulphur	800g/l sulphur	7.6/l
Other		
Calcium	Leaf defence enhancement	2/l
K Repellent	Fly repellent	40/l
Soil Conditioner	(soil enhancement + root stimulant)	6.25/l
Garlic	(plant stimulant) – granules	7/kg
Garlic	(plant stimulant) – liquid	16-20/l
Disinfectant	(controls bacteria, fungi, viruses)	16/l
Black plastic mulch	38 microns gauge	65/200 x 1.3m
	50 microns gauge	200/600 x 1.4m
Biodegradable plastic mulch	18 microns gauge	460/1000 x 1.5m
Woven plastic mulch		55/100 x 1m
Fleece	18 g/m₂²	45/100m x 2m
	25 g/m²	55/100 x 2m
Pest protection netting	17mm mesh anti-bird net	90/100x2m
Birdscarer - humming line		33/500m

Addresses of suppliers of permitted inputs can be obtained from the control bodies. Some products can only be used under restricted circumstances and the certifying body must be consulted prior to their use. Producers should always check the compatibility of new organic pest and disease control products and methods with their control body before use.

Several crop protection inputs require prior approval by the Control Body. Biological control is mainly suitable for use inside, i.e., polytunnels or glasshouses. The only current exception is *Bacillus thuringiensis*, but research into outdoor biocontrol is underway and some growers are working with outdoor biocontrol[10]. However, all these products should be used as a last resort, as they are not free of side-effects and still have the capacity to disrupt beneficial insects and the ecosystem interactions on which the stability of the organic farming system depends. The requirement to use undressed seed in organic farming may increase pest and disease risks at germination, particularly for forage crops, so seed quality and hygiene are very important.

Bi-cropping

Growing two or more combinable crops together increases crop diversity and makes better use of the soil and air environment. There is reduction in pest and disease risk and potential decrease in weeds and where cereals and pulses are grown together an increase in cereal protein and yield, compared to growing the crops separately.

There is potential for bi-cropping of both combinable crops[11], and whole-crop forage[12]. Where grown for crop sale, there is a need to separate cereals from the legumes, at a cost of £8-10/tonne hiring a gravity separator, unless sold for animal feed.

Compatible varieties must be selected. Seed mix depends on use of the crop, but the optimum varies between 50% legume/50% cereal and 80% legume/40% cereal (% of pure stand), sown as a mixture or in alternate rows.

A technical leaflet on growing peas and barley together is available here: https://orgprints.org/id/eprint/43730

Reduced Tillage

Recent interest in reduced tillage and no tillage systems in organic arable cropping is motivated by the aim of improving soil health and claims for significant carbon sequestration. Research indicates that tillage has only a small or no effect on soil carbon levels when monitored through the whole 50cm profile, although reduced tillage may have advantages for soil biology.

10. https://agricology.co.uk/resource/biological-control-strategies-outdoor-vegetable-production/
11. https://agricology.co.uk/sites/default/files/Beans%20and%20wheat%20intercropping.pdf
12. https://tinyurl.com/agricology-bicrop

SHIPTON MILL

The system developed by Rodale using a preceding cover crop such as rye, which is killed using a crimper roller, has proved to be unreliable in UK conditions.

An Innovative Farmers field lab took a different approach and direct drilled a competitive cereal such as oats into white clover, established in the preceding crop. Working with 10 farms over two years in non-replicated trials they found a 30% yield reduction compared to the control, a plough and cultivated treatment. Weed levels were reduced in the second year[13].

Organic Research Centre undertook trials over several years in the Tilman-org project[14] and found yields across 24 sites around Europe were 8% lower than with plough-based cultivations. Using an Eco-Dyn, fixed tine, shallow skimming cultivator, with several passes in the autumn prior to winter wheat yields were substantially lower due to poorer soil structure and restricted N availability.

A number of farmers have successfully managed non-inversion tillage over many years using a variety of fixed and flexible tined cultivators at various depths in order to incorporate a ley and control weeds.

While reduced or no tillage saves fuel, benefits soil life and may result in some relatively small additional carbon sequestration compared to ploughing and cultivation, it may also result in significantly lower yields and some increase in weeds. Shallow ploughing i.e., less than 20cm, remains the best compromise for the majority of organic farms, giving good yield and weed control and higher organic matter than deep ploughing or cultivation[15].

13. https://www.agricology.co.uk/resources/living-mulches-final-report
14. https://www.tilman-org.net/tilman-org-home-news.html
15. https://orgprints.org/id/eprint/29974/

Further reading

Briggs S (2008) *Organic Cereal and Pulse Production: a complete guide.* Crowood Press, Marlborough.

Davies G, Turner B, Bond B (2008) *Weed Management for Organic Farmers, Growers and Smallholders – a complete guide*, Crowood Press, Marlborough.

Davies G, Sumption P, Rosenfeld A (2010) *Pest and Disease Management for Organic Farmers, Growers and Smallholders – a complete guide*, Crowood Press, Marlborough.

Lampkin N (1990) *Organic Farming.* Old Pond (currently out of print).

Sumption P (2023) *The Organic Vegetable Grower - a Practical Guide to Growing for Market.* Crowood Press, Marlborough (in press).

Websites and technical guides

Agricology - resource category for soils and crops: http://www.agricology.co.uk

Organic farm knowledge hub: http://farmknowledge.org/index.php

FIBL/ORC (2016) The Basics of Soil Fertility http://tinyurl.com/ORC-FiBL-soil

Managing Manure on Organic Farms; EFRC & ADAS; http://orgprints.org/24819

ORC Technical leaflets and Research Reviews
www.organicresearchcentre.com/resources/publications

- 4: *Soil analysis and management*
- 5: *Compost - the effect on nutrients, soil health and crop quantity and quality*
- 6: *A guide to nutrient budgeting on organic farms*
- 7: *Composting with rock phosphate*
- 8: *Managing phosphorus dynamics in organic rotations*

IOTA Research Reviews www.organicresearchcentre.com/resources/publications

- Wright I (2008) Combinable protein crop production
- Preston K (2008) Management and sustainability of stockless organic arable and horticultural systems

Innovative Farmers Field labs https://www.innovativefarmers.org

Technical guides produced by the Soil Association https://www.soilassociation.org/farmers-growers/technicalinformation

Organic Farmers & Growers Ltd. https://ofgorganic.org/useful-info/downloads

General information and assumptions

The crops included in this section are combinable crops most commonly grown on organic farms and for which a market exists. Information on potatoes, root crops and field-scale vegetables is shown in Section 8. Forage crops and green manures are in Section 9. Less widely grown crops (sugar beet, oilseed rape, vining peas) have not been included because there is currently no significant market for them due to the lack of processing opportunities.

Yields of organic cereal crops reach about 50-60% of conventional yields, but there is considerable variability in yields depending on soil type, the place in the rotation and crop management. Grain legume yields do not depend on nitrogen availability, but are well known for yield variability, often related to weeds and disease and intolerance to water stress. Yield projections are based on commercial farm data and can be expected under good management on soils well suited to the individual crops, typically grade 1 and 2.

Prices

Marketing and transport costs are not included in these margins, but organic shipments tend to be smaller and travel longer distances than conventional. Transport costs can therefore have a significant impact on price and have increased considerably in recent years. Prices can also vary considerably due to seasonality and market outlet. Unless indicated otherwise, pre-January sale to millers/wholesale grain traders has been assumed. Price assumptions are based on current trading prices with the sensitivity analysis indicating possible trends and ranges.

Quality requirements are important for obtaining a premium for milling markets and, if not met, the crop may have to be sold for feed or at lower prices. Allowance needs to be made for the time and effort required for marketing. The UK is only 20%[1] self-sufficient in organic animal feed stuffs, so UK crops do compete with imports. There is also some export trade in small volumes of malting barley. More collaborative marketing also provides opportunities, for example through Organic Arable and Organic Herd (previously OMSCO) establishing partner schemes for the direct supply of grain and fodder crops between farms on contract.

1. Soil Association *A new era for UK Organic Cereals* 2022 Handbook for Arable Farmers and Advisors https://tinyurl.com/SA-Cereals22

In-conversion feed cereals can be included at up to 30% of livestock rations. They trade at about £20/t less than full organic cereals and can generate some additional cashflow during the conversion period, but the need for fertility building and the potential negative effects on the value of fully organic crops should be considered.

Straw

In principle, straw should not be sold off the holding to avoid losing potassium. Producers selling organic wheat straw for organic mushroom production (used together with poultry litter as compost) should aim for the used compost to be returned. Prices will vary according to locality, species and any agreement for the return of compost/manure. Straw sales are not included in the gross margins presented.

Seed rates, timing and seed prices

Seed rates should be 10-15% higher than standard to compensate for losses due to birds, pests and diseases. They may need to be reduced to minimise competition if the crop is to be undersown with grass and clover. Drilling of cereals may be delayed in autumn to allow for weed germination and control, and to minimise fungal and viral disease carry-over risks, but this should not be at the expense of good drilling conditions. Costs for organic seeds based on the last season have been used in the gross margins for all crops where organic seeds are likely to be available. Home-saved seed should be tested for germination, disease and admix.

Cereal variety selection[2]

Careful selection of varieties is necessary for cereal production (especially wheat) taking into account the market as well as criteria such as crop architecture, height, tillering, disease resistance, and emergence characteristics. The AHDB Recommended Lists for Cereals and Oilseeds https://tinyurl.com/AHDB-recc-lists-cereals provide some information on the characteristics of varieties used in conventional farming, however the trials are all undertaken with standard fertiliser inputs and even the results from untreated trials are of limited value to organic farming. Variety trials under organic conditions are undertaken by the Organic Research Centre and by Organic Arable: https://tinyurl.com/ORC-QUOATS

---

2. Cereal variety and population selection, IOTA PACARes Research Review, https://tinyurl.com/IOTA-cereal-population Pearce, B (2004)

   Organic Cereal Varieties: The results of four years of trials. EFRC technical and Research notes. www.orgprints.org/4139

Newer varieties can be as productive as some of the older varieties. The UK version of the European seed website www.organicxseeds.co.uk provides a database of suppliers of organic seed.

Mixtures of cereal varieties grown in the field can have positive benefits over pure stands in terms of disease and pest control and buffering against environmental variables. They may have a more limited market for human consumption but are suitable as animal feed.

Cultivations and weed control[3]

Normal primary and secondary cultivations and establishment methods are assumed including ploughing, cultivating, drilling, harrowing and rolling. Additional secondary cultivations or a bastard fallow in late summer prior to winter crops may be needed pre-drilling for perennial weed control. Annual weeds may be controlled using false seedbeds and harrowing. Perennial weeds, e.g., couch, can be controlled pre-drilling by frequent shallow cultivations to desiccate the rhizomes. Hand-pulling of wild oats and docks is sometimes resorted to but can be expensive. Inter-row hoeing on a wide row system is increasing in popularity, particularly using vision-guided hoe systems.

Crop nutrition and fertiliser use

Farm-sourced manures will be primarily applied to the ley, particularly where P or K levels are low, but may also be applied to winter crops in spring and prior to spring crops. Although it is not general practice, organic standards permit the use of some brought-in non-organic manures, depending on livestock system origin (including feed sources) and treatment, up to the equivalent of 170kg N/ha. Exchange between organic farms of manures for straw via a 'linked unit' system is also acceptable. Manure and slurry are not classified as waste in the recent regulations, so these arrangements may be practised without a Waste Management Licence. Green waste composts can be an alternative (see Section 12).

In the gross margins lime and rock phosphate are applied on a rotational basis as needed (indicated by soil analysis). A standard charge of £44/ha (£17.80/ac) has been applied to each gross margin on the following basis:

- Phosphate (27%) 0.5t/ha (0.2t/ac) @ £260/t every 7 years = £19/ha (£7.70/ac)
- Lime 5t/ha (2t/ac) @ £35/t every 7 years = £25/ha (£10/ac).

Pest and disease control

3. http://www.gardenorganic.org.uk/weed-management

For most cereal and grain legume crops, rotation design is fundamental to the control of soil-borne pests (in particular nematodes) and diseases (see rotation design above). Aphids can be controlled through habitat management using flowering species such as marigold, ox-eye daisy or Phacelia in field margins as well as 'beetle banks' to attract beneficial insects (see Section 9 for costings).

Disease control is achieved primarily through varietal resistance. Variety or species mixtures may also be an option. Some inputs (e.g., sulphur) are permitted but seldom used for arable crops.

Other variable costs

A standard charge of £40/ha (£16/ac) has been included in each of the gross margins to cover occasional items such as casual labour for hand-rogueing, foliar seaweed feeds and sulphur applications for disease control. Major weed infestations may incur substantially higher costs for hand-rogueing.

Post-harvest management

Investment in grain cleaning, drying and storage facilities may be needed to maintain quality to avoid price penalties. Organic grain is sold on a similar specification as conventional with a maximum of 15% moisture and 2% admixture for feed. Ergot can be a problem in certain areas in some years. Technology based on colour separation can clean grain effectively but at a cost.

Sensitivity analysis

Each gross margin includes a sensitivity analysis of key variables, illustrating the impact of a change in conditions that are likely to have significant influence. The first two columns show the influence on the gross margin of a change (as indicated) in the values of certain parameters (e.g., price, yield). The remaining columns show the range of values which might be found in practice and the resulting range in gross margins.

Yields used in the sensitivity analysis of the gross margins reflect what is achievable across a typical range of conditions of soil type, climate and the crop's place in the rotation. Low yields may represent poor yields on good soil types or moderate/good yields on poorer soil type. The high yields illustrate the potential of the crop on good soils/good position in the rotation. The price

4. ORC Publications https://www.organicresearchcentre.com/resources/publications

   IOTA Technical Leaflets: Composting with rock phosphate (No 7); Managing phosphorus dynamics in organic rotations (No 8). https://tinyurl.com/IOTA-Pdynamics

range is carefully chosen to illustrate the impact of possible changes. Where a cost item is included in the sensitivity analysis, e.g., casual labour, the high gross margin value corresponds to the low input use value.

Further reading

Briggs, S (2008) *Organic Cereal and Pulse Production: a complete guide.* Crowood Press, Marlborough.

IOTA Research Reviews on arable:
https://www.organicresearchcentre.com/resources/publications/

Soil Association *A new era for UK Organic Cereals* 2022 Handbook for Arable Farmers and Advisors https://tinyurl.com/SA-Cereals22

Wheat

Output

Price/ quality Feed prices have been assumed (see sensitivity analysis). A strong feed wheat market exists because of the need to use only organic feed for nearly all livestock. Higher yielding varieties especially for this market can grow on land less suited to milling wheat. The market for milling wheat is limited by the quality of the sample: bread-making specifications are >13% protein at <15% moisture, >250 Hagberg, >76kg/hl specific weight. For each 0.1% fall in protein deduct £1/t down to a min of 11%. High protein levels are difficult to achieve, requiring yield and quality considerations to be balanced. Spring varieties have greater potential achieving milling quality on many sites but are less suited to flaking. Milling Wheat price is currently £330/t. There is a small biscuit wheat and malt wheat market for crops meeting the right specification. Grain should be cleaned before sale. In-conversion £15/t less than full organic.

Straw Yields 3-4t/ha, straw value not included. Thatching straw (e.g., Maris Widgeon) can fetch higher prices. In both cases labour and contractor costs for harvesting, handling, haulage, etc., need to be considered

Variable costs

Seeds Seed rate: winter wheat 400-450 seeds/m^2 = 180-220kg/ha (1.5-1.8cwt/ac); spring wheat 500-550 seeds/m^2 = 225-275kg/ha (1.8-2.2cwt/ac). Higher seed rates may be required due to later drilling (for weed and disease control) and reduced establishment, although too high a seed rate may reduce specific grain weight. Organic seed prices are assumed; early ordering will ensure availability of the desired variety.

Crop management

Varieties Careful selection of varieties is necessary, paying particular attention to protein potential for milling wheat. In very fertile situations, winter wheat generally outperforms spring-sown crops. In less fertile and weedier situations, spring-sown crops can perform just as well.

Rotation First cereal after grass/legume break or residual fertility from crops with manure applications (e.g., potatoes/maize) and where good weed control can be obtained. Should not be grown more than twice in succession. Declining soil N levels reduce tillering and protein level, increased risk of take-all, and weed competition.

Weeds Post-emergence mechanical weed control with harrow-comb type weeder or in wide rows with inter-row hoeing. The number of passes required depends on weed competitiveness and crop density. Spring wheat is a poor competitor resulting in higher cultivation costs. Long-strawed varieties can help with weed suppression. Hand-rogueing of wild oats and docks can be a significant cost.

Harvesting Prompt harvesting and drying (to <15%) to ensure quality and reduce storage problems.

Winter wheat

					£/ha	*(£/ac)*
Grain Feed	4.2 t/ha	*(1.7 t/ac)*	@	295 £/t	1239	*(501)*
Total output					**1239**	*(501)*
Seed	200 kg/ha	*(1.6 cwt/ac)*	@	650 £/t	130	*(53)*
Fertilisers	Applied on rotational basis, see p.103				62	*(25)*
Other	See p. 104				40	*(16)*
Total variable costs					**232**	*(94)*
Gross margin					**1007**	*(408)*

Sensitivity analysis For explanation, see p. 104

	Change in value (+/-)	*Change in gross margin*	*Value range Low*	*High*	*Gross margin range Low*	*High*
Yield	0.5 t/ha	148 *(60)*	3.8	5.5	889 *(360)*	1391 *(563)*
Prices						
Feed, organic	10 £/t	42 *(17)*	265	315	881 *(357)*	1091 *(442)*
Bread, milling	35 £/t	147 *(59)*	300	350	1028 *(416)*	1238 *(501)*
Feed, in-conversion	15.00 £/t	63 *(25)*	300	335	1028 *(416)*	1175 *(476)*
Feed, conventional	-105.00 £/t	-441 -*(178)*	175	205	503 *(204)*	629 *(255)*

Spring wheat

					£/ha	*(£/ac)*
Grain Feed	3.2 t/ha	*(1.3 t/ac)*	@	295 £/t	944	*(382)*
Total output					**944**	*(382)*
Seed	250 kg/ha	*(2.0 cwt/ac)*	@	650 £/t	163	*(66)*
Fertilisers	Applied on rotational basis, see p.103				62	*(25)*
Other	See p. 104				40	*(16)*
Total variable costs					**265**	*(107)*
Gross margin					**680**	*(275)*

Sensitivity analysis For explanation, see p. 104

	Change in value (+/-)	*Change in gross margin*	*Value range Low*	*High*	*Gross margin range Low*	*High*
Yield	0.5 t/ha	148 *(60)*	3.0	4.0	621 *(251)*	916 *(370)*
Prices						
Feed, organic	10 £/t	32 *(13)*	265	315	584 *(236)*	744 *(301)*
Bread, milling	35 £/t	112 *(45)*	300	350	696 *(281)*	856 *(346)*
Feed, in conversion	15.00 £/t	48 *(19)*	300	335	696 *(281)*	808 *(327)*
Feed, conventional	-105.00 £/t	-336 -*(136)*	175	205	296 *(120)*	392 *(158)*

Spelt wheat

Output

Price/output There is a very specific market for spelt products; it provides a good nutritional alternative to those wishing to avoid common wheat. There are only two or three main outlets and volumes are limited. It would be unwise to grow the crop without a contract. Requires dehulling prior to use which results in losses of about 35% of the harvested weigh. Additional costs of haulage to processors must also be factored in and these vary significantly according to quantity and distance. The quality specifications are for 12% protein, 240 Hagberg and 77kg /hl bushel weight. It is not an easy crop to sell on the open market and there are few opportunities for a failed crop not making the milling specifications. However, spelt is apparently an excellent calf feed as the fibre content of the grain is good at stimulating rumen development.

Variable costs

Seeds Seed rate 180kg/ha (1.4cwt/ac). If grown on contract, seed may be supplied; organic seed prices have been assumed.

Crop management

Rotation First or second cereal, autumn sown

Varieties It is important to grow pure spelt varieties for human consumption. Some varieties have been crossed with wheat to improve yield and ease of hulling, but there are concerns that this introduces some of the genes which are responsible for wheat intolerance and so such varieties are not acceptable for this market.

Cultivation The crop is suited to a range of soils, particularly those not suitable for winter wheat. It will perform correspondingly better on good soils. It is a tall, aggressive crop that suppresses weeds and has no significant pest or disease problems, apart from being susceptible to lodging.

Spelt wheat

					£/ha		*(£/ac)*
Grain Milling	3.5 t/ha	*(1.4 t/ac)*	@	370 £/t	1295		*(524)*
Total output						**1295**	*(524)*
Seed	180 kg/ha	*(1.4 cwt/ac)*	@	925 £/t	167		*(67)*
Fertilisers	Applied on rotational basis, see p.103				62		*(25)*
Other	See p. 104				40		*(16)*
Total variable costs						**269**	*(109)*
Gross margin						**1027**	*(415)*

Sensitivity analysis For explanation, see p. 104

	Change in value (+/-)	*Change in gross margin*	*Value range Low*	*Value range High*	*Gross margin range Low*	*Gross margin range High*
Yield	0.5 t/ha	185 *(74)*	2.5	3.7	657 *(266)*	1101 *(445)*
Prices						
Milling, organic	10 £/t	32 *(13)*	340	390	847 *(343)*	1007 *(408)*
Feed, organic	-75 £/t	-240 -*(96)*	230	280	495 *(200)*	655 *(265)*

Barley

Output

Price/ quality The market for organic barley has been more volatile than wheat, but it has been strengthened by increasing demand for feed grains, malting and the major reliance on imports. The demand for feed barley is related to price, supply and demand for feed wheat. Demand for malting barley in the UK is now significant with an expanding number of maltsters and breweries producing organic beer, and there is some export. Malting barley should be grown on contract, but if quality specifications are not met this will have to be sold as feed at a lower price. Spring barley may be more likely to achieve a good malting quality and low soil nitrogen conditions may also be beneficial in this respect. Some barley for whole grain and/or flaking is required by the whole food market, and as an ingredient for muesli, but this demand is also very limited. Price assumed is for feed as the most likely outlet. There is a limited market for in-conversion feed barley.

Straw Yield 2.5-3t/ha. Suitable for feeding to livestock, value not included.

Variable costs

Seeds Seed rate: winter barley 350-400 seeds/m^2 = 160-200kg/ha (1.3-1.6cwt/ac); spring barley 375-425 seeds/m^2 = 180-220kg/ha (1.5-1.8cwt/ac). Variety choice for disease resistance is important. Organic price is assumed for all.

Fertilisers Top-dressing with manure/slurry may be required.

Crop management

Rotation First or second cereal. Spring barley can be undersown with ley mixture. Low soil-N availability in early spring, weed competition and disease susceptibility make winter barley a difficult crop to grow under organic conditions.

Weed control Pre-drilling weed cultivations in the autumn will depend on weed incidence and may not be necessary. Post-emergence mechanical weed control with harrow-comb type weeder, depending on weed incidence. The earlier harvest dates for barley provide a good opportunity for subsequent cultivations to control couch, thistle etc.

Disease Variety selection should be the main approach. Variety mixtures may provide added protection, but suitability depends on target market.

Winter barley

					£/ha		(£/ac)
Grain Feed	3.0 t/ha	(1.2 t/ac)	@	260 £/t	780		(316)
Total output						**780**	(316)
Seed	180 kg/ha	(1.4 cwt/ac)	@	600 £/t	108		(43)
Fertilisers	Applied on rotational basis, see p.103				62		(25)
Other	See p. 104				40		(16)
Total variable costs						**210**	(84)
Gross margin						**570**	(231)

Sensitivity analysis For explanation, see p. 104

	Change in value (+/-)	*Change in gross margin*	*Value range Low*	*High*	*Gross margin range Low*	*High*
Yield	0.5 t/ha	130 (52)	2.5	3.5	440 (178)	700 (283)
Prices						
Malting, organic	25 £/t	75 (30)	255	305	555 (225)	705 (285)
Feed, organic	10 £/t	30 (12)	230	280	480 (194)	630 (255)
Feed, conventional	-100.00 £/t	-300 -(120)	145	165	225 (91)	285 (115)

Spring barley

					£/ha		(£/ac)
Grain Feed	3.2 t/ha	(1.3 t/ac)	@	260 £/t	832		(337)
Total output						**832**	(337)
Seed	230 kg/ha	(1.8 cwt/ac)	@	600 £/t	138		(55)
Fertilisers	Applied on rotational basis, see p.103				62		(25)
Other	See p. 104				40		(16)
Total variable costs						**240**	(96)
Gross margin						**592**	(240)

Sensitivity analysis For explanation, see p. 104

	Change in value (+/-)	*Change in gross margin*	*Value range Low*	*High*	*Gross margin range Low*	*High*
Yield	0.5 t/ha	130 (53)	3.0	4.0	540 (219)	800 (324)
Prices						
Malting, organic	25 £/t	80 (32)	255	305	576 (233)	736 (298)
Feed, organic	10 £/t	32 (13)	230	280	496 (201)	656 (265)
Feed, conventional	-100.00 £/t	-320 -(130)	145	165	224 (91)	288 (117)

Oats

Output

Price/ quality The strong and growing demand for oats for human consumption (flaking) has resulted in a significant increase in production. Contracts are advisable due to a limited number of UK processors. There is also a ruminant feed grain market with price related to the feed wheat market. The price assumed is for flaking oats, for which quality requirements are minimum 50kg/hl, and up to 6% screenings, however it may not be possible to achieve milling quality every year. Appearance is important as samples can be rejected for discoloration. Variety choice is also important; millers look for good kernel content. There is a limited market for in-conversion oats for animal feed.

Straw Yield 3-4t/ha, value not included.

Variable costs

Seeds Seed rate: winter oats 500-550 seeds/m^2 = 175-225kg/ha (1.4-1.8cwt/ac); spring oats 650-700 seeds/m^2 = 220-270kg/ha (1.8-2.2cwt/ac). Estimated prices are for re-cleaned, undressed organic seed.

Crop management

Rotation The ability of oats to compete against weeds and to thrive in lower fertility conditions than wheat make this crop an ideal second or third cereal, however the yield will be significantly reduced if grown in a less fertile position in the rotation or on poorer quality soils. Oats should not be grown more than once in succession because of cereal cyst nematode risk.

Weed control Will depend on weed incidence and may not always be necessary. Both pre-drilling weed cultivations and post-emergence mechanical weed control with harrow-comb type weeder are possible.

Winter oats

					£/ha		(£/ac)
Grain Milling	4.0 t/ha	*(1.6 t/ac)*	@	280 £/t	1120		*(453)*
Total output						1120	*(453)*
Seed	195 kg/ha	*(1.6 cwt/ac)*	@	700 £/t	137		*(55)*
Fertilisers	Applied on rotational basis, see p.103				62		*(25)*
Other	See p. 104				40		*(16)*
Total variable costs						239	*(95)*
Gross margin						882	*(357)*

Sensitivity analysis For explanation, see p. 104

	Change in value (+/-)	*Change in gross margin*	*Value range Low*	*High*	*Gross margin range Low*	*High*
Yield	0.5 t/ha	140 *(56)*	3.5	4.8	742 *(300)*	1106 *(447)*
Prices						
Flaking, organic	10 £/t	40 *(16)*	250	300	762 *(308)*	962 *(389)*
Feed, organic	10 £/t	40 *(16)*	210	260	602 *(243)*	802 *(324)*
Feed, conventional	-80.00 £/t	-450 -*(180)*	150	170	362 *(146)*	442 *(179)*

Spring oats

					£/ha		(£/ac)
Grain Milling	4.3 t/ha	*(1.7 t/ac)*	@	280 £/t	1204		*(487)*
Total output						1204	*(487)*
Seed	220 kg/ha	*(1.8 cwt/ac)*	@	700 £/t	154		*(62)*
Fertilisers	Applied on rotational basis, see p.103				62		*(25)*
Other	See p. 104				40		*(16)*
Total variable costs						256	*(104)*
Gross margin						948	*(384)*

Sensitivity analysis For explanation, see p. 104

	Change in value (+/-)	*Change in gross margin*	*Value range Low*	*High*	*Gross margin range Low*	*High*
Yield	0.5 t/ha	140 *(56)*	3.2	5.0	640 *(259)*	1144 *(463)*
Prices						
Flaking, organic	10 £/t	43 *(17)*	250	300	819 *(331)*	1034 *(418)*
Feed, organic	10 £/t	43 *(17)*	210	260	647 *(262)*	862 *(349)*
Feed, conventional	-80.00 £/t	-344 -*(139)*	150	170	389 *(157)*	475 *(192)*

Output

Price/ quality — The price used is the average for quality milling rye. Contracts are essential, because once the milling market is supplied, the surplus only has a limited value as feed. Milling requirements are normally minimum 180 Hagberg. This can deteriorate rapidly and there is a risk that sample will fail milling specifications. Small mills often have a demand for limited quantities of rye.

Straw — Suitable for bedding, value not included.

Variable costs

Seeds — Seed rate 160-200kg/ha (1.3-1.6cwt/ac). Limited volumes of organic rye seed are available. Estimated prices are from 2022 due to low supply in 2023. Conventional seed prices are at £650/t, but a derogation prior to sowing is required (see Section 3).

Crop management

Rotation — The ability of rye to compete against weeds and to thrive in lower fertility conditions than wheat make this crop a good second or third cereal, although low yield can be a disadvantage (typically 2.5–4.0t/ha). Rye is less palatable to rabbits than other cereals.

Weed control — Pre-drilling weed cultivations in the autumns will depend on weed incidence and may not be necessary. Post-emergence mechanical weed control with harrow-comb type weeder, also not always necessary.

Rye

					£/ha		*(£/ac)*
Grain Milling	3.2 t/ha	*(1.3 t/ac)*	@	310 £/t	992		*(401)*
Total output						992	*(401)*
Seed	185 kg/ha	*(1.5 cwt/ac)*	@	750 £/t	139		*(56)*
Fertilisers	Applied on rotational basis, see p.103				62		*(25)*
Other	See p. 104				40		*(16)*
Total variable costs						241	*(96)*
Gross margin						751	*(304)*

Sensitivity analysis — For explanation, see p. 104

	Change in value (+/-)	*Change in gross margin*	*Value range Low*	*Value range High*	*Gross margin range Low*	*Gross margin range High*
Yield	0.5 t/ha	155 *(62)*	3	3.5	689 *(279)*	844 *(342)*
Prices						
Milling, organic	10 £/t	32 *(13)*	280	330	655 *(265)*	815 *(330)*

Triticale

Output

Price The price is similar to feed barley. Compounders are generally no longer using triticale and the limited market is largely for sale direct to farmers for livestock feed. Yields are typically equal to, or better than, wheat, particularly due to its competitive nature. In-conversion grain may sell for a slightly reduced price than full organic, but the market is limited.

Variable costs

Seeds Seed rate 160-220kg/ha (1.3-1.8cwt/ac). Estimated prices are based on cleaned, undressed, organic seed.

Crop management

Yields Similar to wheat but variable and low yields seen in recent years due to disease (Yellow Rust) and poor germination have resulted in a reduction in the area grown.

Varieties Predominantly winter varieties available. Some winter varieties can be sown up until the end of February.

Rotation Weed competitiveness and tolerance of lower fertility conditions make it suitable as a second or third cereal crop on some farms. Triticale is less palatable to rabbits than other cereals

Weed control Pre-drilling weed cultivations in the autumn will depend on weed incidence and may not be necessary. Post-emergence mechanical weed control with harrow-comb type weeder will depend on weed incidence and may not be necessary.

Triticale

					£/ha		*(£/ac)*
Grain Feed	3.0 t/ha	*(1.2 t/ac)*	@	255 £/t	765		*(310)*
Total output						765	*(310)*
Seed	200 kg/ha	*(1.6 cwt/ac)*	@	570 £/t	114		*(46)*
Fertilisers	Applied on rotational basis, see p.103				62		*(25)*
Other	See p. 104				40		*(16)*
Total variable costs						216	*(86)*
Gross margin						549	*(222)*

Sensitivity analysis For explanation, see p. 104

	Change in value (+/-)	*Change in gross margin*	*Value range Low*	*High*	*Gross margin range Low*	*High*
Yield	0.5 t/ha	128 *(51)*	2.5	3.5	422 *(171)*	677 *(274)*
Prices						
Feed, organic	10 £/t	30 *(12)*	225	275	459 *(186)*	609 *(246)*
Feed, conventional	-75.00 £/t	-225 -*(90)*	165	195	279 *(113)*	369 *(149)*

Field beans[5]

Output

Price Price used is the average wholesale price for field beans, with normal tannin content. Better prices can be achieved for low tannin or white/coloured varieties in the organic market. The premium for organic beans (and peas) is influenced by trading prices for other protein sources (in particular soya). The removal of the allowance to feed 5% conventional feed to monogstrics in 2025 will strengthen the market for protein crops. In-conversion beans may sell for a slightly lower price than full organic.

Variable costs

Seeds Seed rate: winter 25, spring 45 seeds/m^2. Thousand grain weight very variable, so seed rate should be estimated directly each time. Seed prices are for organic seeds.

Other The costs cover occasional foliar sprays of sulphur for mildew, more of a problem in wetter parts of the country, and occasional hand rogueing of weeds.

Crop management

Rotation Break crop after cereals. A non-leguminous green manure can be sown following cereals prior to spring beans.

Cultivation Techniques used vary with conditions (weeds) and seedbed required. Weedy fields require stubble cultivation before ploughing.

Establishment Winter beans may be broadcast and ploughed to 15cm (6") and levelled using a harrow, or drilled as for spring beans. Choice depends on weed control strategy. Drilling with wide-row spacings provides an opportunity for later inter-row cultivations and possibly undersowing, while narrow row spacings or broadcasting can help establish a dense crop to smother weeds.

Weed control A pre-drilling weed strike may be needed depending on weed incidence. Control during initial stages of plant growth using spring-tined harrows or inter-row hoeing. The number of passes will depend on weed competitiveness and crop density. Winter beans can be harrowed hard in the spring and will tiller, providing a good means of weed control. Undersowing is possible and can assist with the control of late emerging weeds.

Pests Flowering field margins, e.g. marigold, ox-eye daisy, Phacelia, are attractive to aphid predators. Spring beans tend to be more susceptible to aphid problems.

Diseases Chocolate spot can be a problem with winter beans and brown rust with spring beans, but there are no direct control measures.

Winter beans

					£/ha		(£/ac)
Grain	3.2 t/ha	*(1.3 t/ac)*	@	450 £/t	1440		*(583)*
Total output						**1440**	*(583)*
Seed	210 kg/ha	*(1.7 cwt/ac)*	@	820 £/t	172		*(69)*
Fertilisers	Applied on rotational basis, see p.103				62		*(25)*
Other	See p. 104	and notes			50		*(20)*
Total variable costs						**284**	*(114)*
Gross margin						**1156**	*(468)*

Sensitivity analysis For explanation, see p. 104

	Change in value (+/-)	*Change in gross margin*	*Value range Low*	*High*	*Gross margin range Low*	*High*
Yield	0.5 t/ha	225 *(90)*	2.90	3.50	1021 *(413)*	1291 *(522)*
Prices						
Feed, organic	10 £/t	32 *(13)*	420	470	1060 *(429)*	1220 *(494)*
Feed, conventional	-220.00 £/t	-704 *-(282)*	215	240	404 *(163)*	484 *(196)*

Spring beans

					£/ha		(£/ac)
Grain	2.7 t/ha	*(1.1 t/ac)*	@	450 £/t	1215		*(492)*
Total output						**1215**	*(492)*
Seed	250 kg/ha	*(2.0 cwt/ac)*	@	820 £/t	205		*(83)*
Fertilisers	Applied on rotational basis, see p.103				62		*(25)*
Other	See p. 104	and notes			50		*(20)*
Total variable costs						**317**	*(128)*
Gross margin						**898**	*(363)*

Sensitivity analysis For explanation, see p. 104

	Change in value (+/-)	*Change in gross margin*	*Value range Low*	*High*	*Gross margin range Low*	*High*
Yield	0.5 t/ha	225 *(90)*	2.5	3.5	808 *(327)*	1258 *(509)*
Prices						
Feed, organic	10 £/t	27 *(11)*	420	470	817 *(331)*	952 *(385)*
Feed, conventional	-220.00 £/t	-594 *-(238)*	215	240	264 *(107)*	331 *(134)*

5. Wright, I (2008) *Combinable protein crop production*. IOTA Res Review. http://orgprints.org/5582/

Dry peas

Output

Price Prices are for animal feed because of the higher feed value and low tannin content, likely to achieve better wholesale price than beans. The increasing requirements for organic feed will improve the market for protein crops. Small market for human consumption, which will trade for a higher price than feed. The market is constrained by the limited number of processing facilities.

Variable costs

Seeds High seed rates should be used to compensate for damage from birds and for rapid canopy cover to assist with weed control. Seed rates vary according to seed size, aim for 60 to 80 seeds/m^2 (= approximately 275-325kg/ha or 2.2-2.6cwt/ac) on medium soils; lower rates (250-285kg/ha or 2.0-2.3cwt/ac) on light soils. Seed prices are for organic seeds. Bird damage can be a significant problem and precautions should be taken. Variety selection is important. Taller crops with a more structured canopy and fuller leaves tend to assist with weed competition. Shorter, semi-leafless varieties are not as competitive against weeds.

Other Costs cover hand rogueing of weeds and occasional application of permitted products for aphid control.

Crop management

Rotation Peas prefer free-draining soils. The position in the rotation should be chosen depending on weeds and nutrient supply; can be used as a break crop after cereals, but fields with high weed pressure should be avoided as peas are not competitive during early and late stages of development. They can be followed by a winter cereal or over-wintering green manure, as they clear the field early.

Cultivation Good soil structure is particularly important. Normally, post-Christmas ploughing is followed by power harrowing and other cultivations to create a fine seedbed.

Establishment Drilled at approx 4-6cm (2") deep. Adequate soil temperature (min 5°C) for rapid emergence is beneficial. Delayed drilling on some soils may be required

Weed control Dry peas are very susceptible to competition from weeds so should only be grown on relatively clean fields. A false seedbed and weed strike greatly assist weed control. Control during initial stages of plant growth using spring-tined harrows or inter-row hoeing (depending on row width) is important, but no passes from shortly before germination until 3-leaf stage to avoid crop damage. The number of passes will depend on weed competitiveness and crop density.

Pests Peas can suffer from aphids. Flowering field margins, e.g., Phacelia, are attractive to aphid predators. Low-level pea and bean weevil attack is common. Given good establishment conditions, the crop will normally grow away from an attack.

Diseases There can be a problem with downy mildew, but there are no effective direct control measures. To prevent the build-up of soil-borne diseases an interval of 6-7 years between pea crops should be respected.

Dry peas

					£/ha		(£/ac)
Grain	3 t/ha	*(1.2 t/ac)*	@	475 £/t	1425		*(577)*
Total output						**1425**	*(577)*
Seed	300 kg/ha	*(2.4 cwt/ac)*	@	820 £/t	246		*(98)*
Fertilisers	Applied on rotational basis, see p.103				62		*(25)*
Other	See p. 104	and notes			50		*(20)*
Total variable costs						**358**	*(143)*
Gross margin						**1067**	*(432)*

Sensitivity analysis For explanation, see p. 104

	Change in value (+/-)	*Change in gross margin*	*Value range Low*	*High*	*Gross margin range Low*	*High*
Yield	0.5 t/ha	238 *(95)*	2.5	4	830 *(336)*	1542 *(624)*
Prices						
Feed, organic	10 £/t	30 *(12)*	445	495	977 *(395)*	1127 *(456)*
Feed, conventional	-250.00 £/t	-750 -*(300)*	215	230	287 *(116)*	332 *(134)*
Human consumption, org.	25 £/t	75 *(30)*	470	550	1052 *(426)*	1292 *(523)*

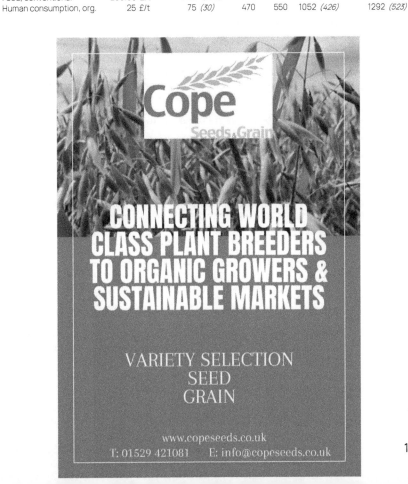

Lupins

Output

Price High livestock feed values (32-40% protein) with a market for both organic and in-conversion crop.

Variable costs

Seeds High seed rates to compensate for pest damage (birds, rabbits) and to ensure rapid canopy cover for weed suppression. Seed rates of 185-225kg/ha (1.5-1.8 cwt/ac). Seed prices assume organic seeds including rhizobial inoculation, which is essential.

Other Hand rogueing weeds and occasional permitted pest and disease control.

Crop management

Varieties Blue, yellow and white lupin varieties with low alkaloids and tannins. Whites are taller, more competitive and have better yield potential on good soils. Yellow and blue varieties are shorter, have more open canopies and are better suited to poor/acid soils. Winter-sown varieties are not well suited to organic systems. Seed is also available as a mixture with triticale.

Soils Lupins prefer acid, free-draining soils between pH 5.0 and pH 7.0. It is important to check the level of free calcium in the soil, as high levels can severely restrict the performance of lupins.

Cultivation Early seedbed preparation is needed to allow weed control prior to establishment (March/April).

Establishment Rhizobial inoculation standard for N fixation (included with seeds). Drilled at approximately 4 cm (1.5"), a good seedbed, early March to mid-April (frost hardy), will establish in poorer seedbeds.

Weed control Lupins are very susceptible to early weed competition and should only be grown on clean fields; avoid fields with deadly nightshade, use stale seedbeds and weed strikes prior to drilling. Control during initial stages of plant growth using spring-tined harrows and/or inter-row hoe.

Pests There is a major threat to establishment from birds and rabbits, so careful and timely precautions are needed. The crops are most under threat at emergence. Low-level pea and bean weevil attack is common but crops likely to grow away.

Diseases Low risk.

Harvest Maturity reached approximately 150 days, harvest in mid-August. Harvested dry, dried to 14% m.c. and stored at a low temp. Can be harvested at 30% moisture content for crimping.

Other Lupins can be grown throughout the UK where soil and climatic conditions are appropriate but organic lupins have proved to be unreliable due to weed competition and low yields. Growing north of the Midlands or in the West may not be advisable due to the late harvest associated with wetter conditions.

Lupins

					£/ha		(£/ac)
Grain	2.5 t/ha	(1.0 t/ac)	@	480 £/t	1200		(486)
Total output						**1200**	(486)
Seed (innoculated)	150 kg/ha	(1.2 cwt/ac)	@	1360 £/t	204		(82)
Fertilisers	Applied on rotational basis, see p.103				62		(25)
Other	See p. 104	and notes			50		(20)
Total variable costs						**316**	(126)
Gross margin						**884**	(358)

Sensitivity analysis For explanation, see p. 104

	Change in value (+/-)	*Change in gross margin*		*Value range Low*	*High*	*Gross margin range Low*		*High*	
Yield	0.5 t/ha	240	(96)	2.0	3.5	644	(261)	1364	(552)
Prices									
Feed, organic	10 £/t	25	(10)	450	500	809	(327)	934	(378)
Feed, conventional	-130.00 £/t	-325	-(130)	320	360	484	(196)	868	(351)
Seed rate	10 kg/ha	13.6	(5.4)	100	190	830	(336)	952	(385)

General

The financial returns to root crops and horticultural enterprises are notoriously variable due to wide variations in yield, price, labour use and marketing costs. Variability has been exacerbated as a result of climate change, with extreme weather events becoming more frequent. The gross margins presented here should only be seen as indicative and must be tailored to individual circumstances. All gross margins shown are based upon returns both from packers and wholesale markets, but do not reflect possible returns from direct marketing to consumers, such as box schemes. At the end of this section, there is some information on performance of small organic production holdings.

This section provides information on yields and prices through various markets, including direct marketing. In organic vegetable rotations some premium crops will be offset by lower or negative gross margins from cereals and fertility-building crops. Returns to some typical vegetable rotations including the potential implications of subsidies are illustrated in Section 5.

Yields

Yield variability may be due to soil type, place in rotation, weather conditions, weed competition and pest and disease damage. In some cases, action can be taken, but usually at additional cost, e.g., irrigation or additional casual labour for weed control. The sensitivity analysis shows yield variability for each crop. In some cases, complete crop failure must be considered a possibility and abandoning a crop may be the best alternative to avoid spiralling costs. We have assumed that most crops are sold at harvest with the exception of top fruit, and that low-cost storage and minimal packaging (bulk containers, boxes or sacks) is used. More sophisticated storage methods can be used for some crops but they add significant cost, so a clear price or market access advantage will need to be identified. For some crops, specific quality or price considerations can make lower yields or an earlier harvest advantageous. Consideration should always be given to the price/yield/quality trade-off for individual crops.

Crop nutrition

A large part of the nutritional requirements of organic vegetable crops, particularly nitrogen, should be supplied by growing fertility-building crops or green manures (cover crops) within the rotation (see also Sections 5). Typically,

these will account for 20 to 40% of field-scale vegetable rotations. Shorter-term green manures can also play a role in retaining fertility and maintaining soil structure. Over-wintering green manure contributes less to N fixation but can play an important role in preventing fertility losses (see Section 9 for seed and other costs of green manure crops). Many farms will also have their own livestock and recycle nutrients through use of own manure or compost.

A standard allowance of £62/ha has been allowed to cover fertilisers applied on a rotational basis (see p. 103) and some use of additional potash sources (such as bought-in manures or potassium sulphate) as required. Purchased farm-yard manure is the most common source of nutrients (provided it is composted before application) but should be checked for potential contamination from herbicides containing aminopyralid and clopyralid, which have caused serious problems for some growers[1]. Other purchased organic and mineral fertilisers used only on an occasional basis have been allowed for under 'other' costs. Please note that mineral fertilisers should only be applied where deficiencies have been identified by soil analyses and in line with the general principles of organic farming. It is recommended to consult the control body prior to their use (see Section 6 for list of inputs and 2023 price ranges).

Under certain circumstances suitable green waste compost can be used. This must be source-separated from a monitored and approved collection system and come within limits of heavy metal concentrations. Certifying bodies should be consulted prior to use and the compost should be produced to Publicly Available Specification (PAS) 100. More information on compost suppliers, compost protocols and research is available from the Compost Certification Scheme website[2]. Green compost prices in 2023 ranged between £0-25/t excluding VAT and transport[3]. Total cost, including purchase price, haulage and spreading can vary significantly between suppliers. WRAP[4] found that the purchase price made up 20-33% of the total cost, haulage between 25-70% and spreading 20-45%.

1. Compost: the effect on nutrients, soil health and crop quantity and quality. IOTA Technical Leaflet 5: https://tinyurl.com/IOTA-compost
2. Compost Certification Scheme www.qualitycompost.org.uk/, also visit: www.biofertiliser.org.uk/standards for further guidance, protocols, and research.
3. https://www.letsrecycle.com/prices/composting/
4. www.wrap.org.uk/
5. IOTA Research Review: Controlling pests and diseases in organic field vegetables, available at: https://tinyurl.com/IOTA-PandD

Crop protection[5]

For root and vegetable crops, the rotation is fundamental to the control of pests and diseases (see Section 6). There should be an emphasis on preventative measures including varietal choice and functional biodiversity (flower strips, beetle banks, companion cropping etc.). Where crop protection inputs are assumed, associated costs are identified specifically (e.g., blight control, biological pest control: see Section 6 for permitted inputs).

Control bodies differ in their approach as to whether permission should be asked before application of products- e.g., the Soil Association insists on submission of a detailed plan and use should be only if there is 'a severe threat to crops'. Copper is still permitted under the retained EU Regulation, but there are currently no copper products licensed for use on crops in the UK. Please check the pesticide register for the authorisation of product and any application limits prior to use. For some inputs, e.g., Bacillus thuringiensis, correct timing is essential to minimise use and maximise effect. Wherever possible other control strategies should be considered. Crop protection inputs used only on an occasional basis are included in 'other' costs.

For all brassica crops, carrots, and in some areas leeks, the use of netting (e.g., enviromesh) for pest control should be considered. Many growers have found that, although expensive, netting can be used on a wide range of crops to control cabbage root fly, carrot fly, flea beetle and aphids, and in addition keeps pigeons, rabbits and other less obvious pests out as well as providing some protection for early crops. Costs are in the range of £1100-1300/ha for the cover and use on approximately 10 different crops plus estimated £320/ha for putting on and taking off for weeding.

Aphids should be controlled by encouraging natural predators through flowering field margins and specific plantings of attractant plants. Monitoring with traps and/or forecasting schemes can aid decisions. The use of potassium soap and natural insecticides should be seen only as a last resort.

Varieties

Variety selection is important with regard to market quality and price, for pest and disease resistance and for weed competitiveness in organic systems. SeedLinked, an innovative online platform and application designed to seamlessly connect growers, facilitate seed sourcing, enable variety rating and comparison, and conduct trials, has recently been launched in the UK and Ireland[6].

---

6. UK users: https://app.seedlinked.com/en-GB/register
   Users in Ireland: https://app.seedlinked.com/en-IE/register

Seeds

Organic seed should be used where possible and prices for organic seed have been used in the gross margins. Growers are allowed to request authorisation of the use of non-organic seed materials or propagating material if no organic material is available. This has increasingly been the case as domestic organic seed production has seen significant decline since the peak of organic production in 2008. A database of organic seed availability (www.organicxseeds.co.uk) must be used to support operators in determining whether organic seed and seed potatoes are available. Operators must have the agreement of their control body for the use of non-organic seed and seed potatoes.

Transplants[7]

All transplants must be organically grown using approved substrates and permitted inputs. Costs assumed include the use of organic seeds. The transplant price is influenced by prices for organic seeds, the use of FI hybrid seed is assumed in cases where normally used (open pollinated seeds would be half the price). The price of transplants varies considerably according to variety, distance from plant-raiser, and the size of order. Costs of plant raising vary throughout the season, with early production being more expensive. We have assumed mid-season or maincrop prices for orders over 4,000.

Labour and machinery

The labour required for horticultural enterprises varies considerably depending on level of mechanisation, timeliness, and the impact of weather conditions on crop and weed growth. Typical ranges are illustrated in the sensitivity analysis for each crop. Standard mechanisation for planting, weed control (including brush, finger or flame weeding), as well as standard (basic) cultivation, ridging, bed construction and harvesting equipment for root crops and onions is assumed (see Section 12). Other machinery costs are not included, as these are assumed to be part of the fixed costs of the holding. These can vary considerably, depending on usage and age of equipment. Machinery costs may be higher on organic horticultural holdings because a wider range of specialist equipment is required, but per hectare values are likely to be lower because of the more extensive nature of the business that includes fertility crops. Labour and machinery costs of a sample of existing organic holdings are illustrated in Section 5.

Labour involved in planting, weed control and harvesting/grading activities is assumed to be carried out by additional, non-regular labour. The casual wage

---

7. *Organic plant raising.* IOTA Research Review: https://tinyurl.com/IOTA-plantraising

rate of £14/h used in the margins was decided to reflect the reality of current wages for jobs not carried out by (for example) a head grower. Employers need to be aware of costs of an additional 13% for National Insurance and employers' liability, and a further 20% for annual and sick leave in in addition to the minimum wage. There may well be scope to reduce these costs through the use of regular labour, increased mechanisation (e.g., weeding) and automated harvesting (e.g., rig harvesters).

Storage and the cool chain

With increased supplies on the organic market, having access to storage facilities is becoming increasingly important for most crops. Cold storage is an important part of the modern production of vegetables and salad crops. Leaf crops especially require the field heat to be quickly removed. However, equipment such as vacuum and ice coolers are expensive and often only feasible for larger scale growers or groups of growers. The costs of cold storage have increased significantly and are expected to increase further due to high energy prices.

Prices and marketing outlets/costs

Apart from crops for industrial processing, prices can show very high seasonal variations and are related to market outlet and quality requirements. In this section, we have assumed field scale production, marketed to pack houses in bulk containers. Price variability is indicated in the sensitivity analysis for each crop. The additional costs and returns of using wholesale markets are shown for comparison for most crops. Intensive market garden operations (including box schemes, see also Section 2) may obtain higher marketable yields and/or prices through direct sales to consumers and retailers, but the added labour, packaging and transport costs need to be considered carefully. The very wide range of traditionally grown crops may also restrict opportunities for mechanisation and specialisation, with subsequent implications for labour requirements and costs.

Comparison of marketing outlets for vegetables

	Pre-pack	Wholesale	Direct Marketing
Marketable yield	Lowest (50-70%)	Medium/high (70-85%)	High (90-95%)
Grade out	30-50%	15-20%	5-10%
Class	Class I (& II)	Class II	Class I & II
Crops range	Limited range	Medium range	High diversity

Pre-packing and wholesale markets

It is essential to secure a market outlet prior to growing a crop. Growing on speculation is likely to disrupt the market and carries a high risk to the grower. Packhouse prices are generally lower than wholesale, but these are often the only realistic outlet for large volumes of crop. Increasingly, the size of the larger box schemes allows them to handle significant volumes. Sales to packers may require less on-farm storage, packing facilities and lower grading costs. However, the more exacting quality requirements for pre-packing are likely to lead to higher growing and harvesting costs and a lower marketable yield than for wholesale. Grading is often carried out at harvest with only Class I crop being taken. Unless indicated otherwise, pre-pack prices are assumed.

Wholesale outlets can generally cope with smaller quantities and lower quality requirements (Class II), resulting in a higher marketable yield, and these may be more suited to medium and smaller operators. Facilities for on-farm grading and packaging are normally required and these carry an additional cost; additional transport costs and marketing commission (assumed at 15% of output) may also be incurred. There may be a need to grow a larger number of crops to secure access to a wholesaler, with negative impacts for economies of scale and labour-saving equipment. For crops that are commonly pre-packed on farms costs are included in the pre-pack section (e.g., cutting and packing cabbage). In all cases, certification charges will be incurred (see Section 3) and these are not included in crop gross margins.

Processing

There is a demand for root crops, vegetables and fruit grown on contract to supply processors for prepared and frozen foods, baby foods, yoghurts, fruit juices and spreads. However, the growth of UK supplies for this market has been slower than expected and prices are often low. Grade-outs from other outlets are by and large not suitable for processing as the quality requirements are quite specific (apart from top fruit for juice making). Certain crops, such as beetroots and strawberries have been included with processing outlets in mind and in top fruit mixed marketing is common. For some crops, growing for processing may require access to freezing and/or storage facilities if the processor is unwilling or unable to take on a full year's supply at one time. Conversely a processor may want to take in a year's supply in one go because of the cost of cleaning down a line for the processing of organic produce. This can have implications for the ability of a grower to supply the required quantity. Freezing and storage costs have not been included throughout, but indications of likely costs have been included in Section 12. The sensitivity analyses include prices for other outlets where available.

Box schemes and direct marketing

Box schemes (see also Section 2) have proved to be a very successful form of marketing for many growers. The aim is to deliver high quality, affordable organic produce to the consumer with a reasonable financial return to the growers. Various types exist including producer-led, where a grower markets mostly his/her own produce; consumer-led, where a group of consumers seek out one or more sources of suitable produce, often buying in bulk and packing themselves and national schemes, some working directly with growers. In Community Supported Agriculture (CSA) schemes producers and consumers work together. Most box schemes consist of a fixed-price box for the consumer, with contents varying from week to week, according to what is in season. The choice of what is packed is left largely to the packer, but increasingly schemes also offer the consumer choice.

Grower-led schemes have the potential for good quality and freshness as all growing and harvesting factors are under direct control. A very wide variety of vegetables is required, creating a heavy workload and the managerial input should not be under-estimated. They may have to buy in produce for continuity, with substantial financial impact. Consumer-led schemes exist but are likely to be vulnerable to crop failures if an alternative source of supply cannot be found. Such problems can be overcome by entrepreneurial schemes involving several suppliers, so that a more consistent supply can be assured. However, the whole supply from one individual holding may not be absorbed by 'entrepreneurial' schemes.

Essential elements of box schemes include the close relationship between producer and consumer, fostering customer loyalty, a stable customer base, financial security and job satisfaction, which is less easy to develop where a third party is involved. The long-term success of box schemes will depend on continued development of this producer/consumer relationship as well as keeping the costs under constant review, especially regarding transport and delivery. It is not enough just to maintain schemes which are working at the moment; continuous development is essential for the future. The economic prospects for box schemes have been impacted by fluctuations in the wider economy and this should be very carefully considered before setting up a new venture.

CSA schemes are increasing, with membership of the CSA Network UK having doubled in the last couple of years. It is a resilient model and attractive to new entrants. A key difference to the box schemes is that people become members and share the risks and rewards of farming with the grower or farmer, committing to buying a share of the harvest for the whole season. Members might also volunteer on the holding on set days. They are essentially non-profitmaking and payment in advance allows greater financial stability and more secure forward planning.

Stallholders at farmers' markets must be producers, usually within 30 miles of the market and sell their own produce. Many markets are not regular enough to provide the continuity required by consumers for vegetables. Producers may use a number of markets in their area. The last few years have seen the emergence of online food hubs and virtual farmers' markets. Such enterprises have been able to capitalise on the rise in home delivery services during the Covid-19 pandemic. Individual crop costings as shown in this section may not reflect returns for box schemes or direct marketing.

Added value and packaging

Higher prices can be obtained through adding value on the farm, in the form of storage (to benefit from higher prices later in the season) or some form of additional processing or packaging. Activities involving processing are subject to a separate licence as the packing of vegetable boxes out of sight of the final customer is defined as 'on-farm' processing. The additional costs and effort associated with these added-value activities need to be carefully considered.

The Soil Association introduced packaging standards in January 2008 in order to minimise the direct and indirect environmental impact of packaging by minimising the amount of material used, maximising the amount of material that can be reused or recycled, and using materials with recycled content wherever possible. Licensees must be able to show at their inspection that they have observed these standards for each packaging type they use. Certain materials must not be used, including PVC, coatings, dyes or inks containing phthalates and wood treated with preservatives. Please check with the certification bodies for their requirements in this area.

Transport

Organic crops may need to travel longer distances to packers and wholesalers than in conventional production. Transport costs have been estimated for most crops at £56/t for bulk transport to the pack-house and transport to wholesalers at approx. £94/t, unless specialist transport charges are likely. It has to be recognised in a period of rapidly changing oil prices that these may change. Where boxes can be re-used several times during a season, and transport distances are shorter, these costs will be reduced correspondingly. Bulkier and more sensitive crops, such as brassicas and salads, will have correspondingly higher transport and container costs.

Profitability of horticultural rotations

As the sensitivity analysis indicates, many horticultural crops are only profitable if the performance is good (high yield and high price, low labour and marketing costs), which requires that production and marketing are kept under constant

review. The potential returns can be high, but it is likely that in any one year on average 10-20% (more for some crops like lettuce) of produce grown will not be marketed. In field-scale vegetable rotations premium crops have to compensate for lower or negative gross margin from cereals or fertility-building crops. The average whole farm gross margins of three rotations are illustrated in Section 5 and are based on individual gross margins of this handbook alongside survey results for organic horticultural holdings.

Support scheme payments

For the new SFI scheme being rolled out through 2023 and 2024 in England, there will be no minimum and maximum land area for grant applications. A minimum size of 3 hectares is currently proposed for eligibility under the Sustainable Farming Scheme in Wales, although the scheme is under development and other provisions for small scale holdings have been proposed. The Scottish Government has also removed its previous minimum threshold of 3 hectares for organic farmers only.

Further reading

Collyns K (2013) *Gardening for Profit – From home plot to market garden* Green Books.

Cubison S (2009) *Organic Fruit Production and Viticulture*. Crowood Press, Marlborough.

Davies G, Lennartson M (eds.) (2005) *Organic Vegetable Production: A Complete Guide*. Crowood Press, Marlborough. (eBook only)

Hall J, Tolhurst I (2006) *Growing Green*. Vegan Organic Network.

Little T, Frost D (2008) *A Farmer's Guide to Organic Fruit and Vegetable Production*. Technical Guide, Organic Centre Wales: Aberystwyth.

Davies, G, Turner, B and Bond, B (2008) *Weed Management for Organic Farmers, Growers and Smallholders: A Complete Guide*. Crowood Press. Marlborough.

Davies G, Sumption P, Rosenfeld A (2010) *Pest and Disease Management for Organic Farmers, Growers and Smallholders – A Complete Guide*. Crowood Press. Marlborough

Bevan et al (1997) *Storage of Organically Produced Crops* Report for Defra https://orgprints.org/8241/

IOTA Research Reviews, *Organic plant raising* https://tinyurl.com/IOTA-plantraising & *Controlling pests and diseases in organic field vegetables*: https://tinyurl.com/IOTA-PandD

Sumption P (2023) The Organic Vegetable Grower- a practical guide to growing for the market Crowood Press, Marlborough (in press)

Technical information, and back copies of *The Organic Grower* magazine (indexed), are available in the members' area of the Organic Growers Alliance (OGA). www.organicgrowersalliance.co.uk

Seed potatoes

Output

Price Sale of seed (74%) to seed merchant and ware (26%) assumed. The market for seed potatoes has shrunk significantly over the last decade, as has the the volume of seed potatoes produced domestically.

Yield Varies with soil type and variety 15-25 t/ha; total 19t/ha assumed

Variable costs

Seeds Rates vary between 4 and 5t/ha depending on tuber seed size. Organic seed should be used where available, non-Organic C1 seed price has been assumed, derogation required. The price of seed for seed potato depends significantly on its generation. Generation 4 or 5 is assumed. For a registered seed company, there are additional marketing licence fees of between £50 and £80 per tonne to consider.

Other See notes for potatoes .

Crop management

As for ware potatoes with a minimum 5 year gap between potatoes.
Production on high ground (over 200m) is preferable to avoid aphid transmitted virus.
Varietal purity and freedom from seed-borne disease such as rhizoctonia is essential.

Seed potatoes					£/ha		*(£/ac)*
Sale seed	14 t/ha	*(5.7 t/ac)*	@	700 £/t	9975		
Sale of ware	5 t/ha	*(1.9 t/ac)*	@	400 £/t	1900		
Marketable yield	19 t/ha	*(7.6 t/ac)*	@	625 £/t	11875		*(4806)*
Output	75% seed	25% ware	Gradeout	15% farm		11875	*(4806)*
Seed	4.3 t/ha	*(1.7 t/ac)*	@	850 £/t	3655		*(1479)*
Fertilisers	Standard rotational with additional potash/FYM				82		*(33)*
Weed control	2 ridge-up + knock down		@	93 £/ha	186		*(75)*
Haulm removal gas					80		*(32)*
Casual labour - planting	25 h/ha	*(10 h/ac)*	@	14.00 £/h	350		*(142)*
- hand rogueing	6.5 h/ha	*(3 h/ac)*	@	14.00 £/h	91		*(37)*
- harvest	24 h/ha	*(10 h/ac)*	@	14.00 £/h	336		*(136)*
- grading	22 t/ha	2 h/t	@	14.00 £/h	626		*(253)*
Seed royalties and certification					330		*(134)*
Defra inspection					103		*(42)*
Transport/bulk	19 t/ha	*(7.6 t/ac)*	@	56 £/t	1064		*(431)*
Other	See notes				50		*(20)*
Total variable costs						6953	*(2814)*
Gross margin						4922	*(1992)*

Sensitivity analysis	*Change in value (+/-)*	*Change in gross margin*	*Value range Low*	*High*	*Gross margin range Low*	*High*
Marketable yield	1 t/ha	541 *(216)*	15	25	2758 *(1116)*	8168 *(3306)*
Price	10 £/t	143 *(57)*	600	1000	3497 *(1415)*	9197 *(3722)*
Casual labour	10 h/ha	140.0 *(56)*	50	150	4225 *(1710)*	5625 *(2276)*

Potatoes

Output

Price Price varies considerably depending on overall production, quality, season, outlet and national production. Limited markets for earlies and for processing.

Yield Varies with water availability (rainfall & irrigation), place in rotation, fertility level soil type and quality (incidence of scab, splits, blight, skin finish, size) and outlet.

Variable costs

Seeds Rates vary between 4 and 5t/ha depending on tuber seed size. Organic seed should be used where available, non-Organic C1 seed price has been assumed, derogation required. The price of seed for seed potato depends significantly on its generation. Generation 4 or 5 is assumed.

Varieties Should have rapid establishment, weed suppression and early tuber bulking, resistance to pests and diseases, in particular late blight[8]. The range of varieties with good blight resistance is increasing, with some retailers signing up to the 'Robust potato Pledge'[9].

Irrigation Will probably be needed to achieve the required skin finish, yield and tuber size. Overall cost of applying 25 mm/ha can range from £90-170/ha.

Harvest Assumed mechanical harvest, carted, clamped, riddled & loaded into bulk containers for pre-packer. Contract harvest & cart £1,055/ha; 0.75-2ha/8-hour-day

Other De-stoning is commonly required, contract charge £285 - 390/ha. Occasional hand rogueing, foliar dressings of seaweed.

Crop management[10]

Rotation After cereals or grass/legume ley (wireworm risk for leys over 2 years old), followed by cereals or brassicas benefiting from residual value of manure.

Establish-ment Conventional, primary and secondary cultivations, ridging and de-stoning (if required). Manual/automatic operated 2 or 3-row planter

Nutrition Farm sourced manure at 25 t/ha. Avoid applying lime before potatoes to avoid scab. May need potassium sulphate on lighter soils (prior approval required).

Weed control 10 days after planting using harrows, ridgers or purpose-built weeders, number of passes depends on weed competition. Inter-row cultivation post-emergence. Pre-emergence flame weeding possible but expensive and not usually necessary. On a small-scale mulching with fresh grass/clover clippings applied with muck spreader is an option to reduce tillage and conserve moisture.

Disease control Blight control through preventive strategies including chitting, use of early-maturing, blight-resistant varieties[10], mixtures; haulm removal, delayed harvest. Refuse piles are seen as the single most important source of infection. Irrigation will reduce common scab incidence, but can increase the incidence of blight, trickle irrigation is most suitable. Other skin blemishing diseases such as stem canker, black surf and powdery scab and viruses can be reduced by using clean seed.

Haulm removal Assumed topping, but commonly also through burning-off.

Storage Suitable facilities will be required.

---

8. https://potatoes.agricrops.org/varieties
9. www.potatodiversitylab.co.uk/about
10. *Organic Potatoes: Cultivating quality – step by step.* FiBL & ORC

Maincrop potatoes

					£/ha		(£/ac)
Marketable yield	23 t/ha	(9.2 t/ac)	@	400 £/t	9200		(3723)
Output	70% of gross yield. Gradeout:			15% farm	15% packer	**9200**	(3723)
Seed	2.5 t/ha	(1.0 t/ac)	@	780 £/t	1950		(789)
Fertilisers	Standard rotational with additional potash/FYM				82		(33)
Irrigation	75 mm		@	4.8 £/mm	360		(146)
Weed control	2 ridge-up & knock down		@	93 £/ha	186		(75)
Casual labour - planting	25 h/ha	(10 h/ac)	@	14.00 £/h	350		(142)
- harvest & grade	33 t/ha	(13 t/ac)	@	31 £/t	1016		(411)
Transport	28 t/ha	(11.2 t/ac)	@	56 £/t	1571		(636)
Haulm removal	1 topping		@	45 £/ha	45		(18)
Other	See notes				50		(20)
Total variable costs						**5610**	(2270)
Gross margin						**3590**	(1453)
Adjustment for wholesale sales							
Marketable yield/output	28 t/ha	(11.2 t/ac)	@	480 £/t	13464		(5386)
less commission @	15%				2019.6		(808)
Additional casual labour	28 t/ha	1 h/t	@	14 £/t	393		(157)
Packaging	28 t/ha	40 bags/t	@	60 p/bag	673		(272)
Additional transport	28 t/ha	(11.2 t/ac)	@	38 £/t	1066		(431)
Gross margin (wholesale)						3702	(1481)

Sensitivity analysis	*Change in value (+/-)*	*Change in gross margin*	*Value range Low*	*High*	*Gross margin range Low*	*High*
Marketable yield	1 t/ha	300 *(120)*	15	40	1190 *(482)*	8690 *(3517)*
Packer price	10 £/t	230 *(92)*	350	450	2440 *(987)*	4740 *(1918)*
Casual labour - additional	10 h/ha	140.0 *(56)*	50	150	2856 *(1156)*	4256 *(1722)*

Early potatoes

					£/ha		(£/ac)
Marketable yield	12 t/ha	(4.8 t/ac)	@	650 £/t	7800		(3157)
Total output	85% of gross yield. Gradeout:			5% farm	10% packer	**7800**	(3157)
Seed/chitting	3.0 t/ha	(1.2 t/ac)	@	780 £/t	2340		(947)
Fertilisers	Standard rotational with additional potash/FYM				82		(33)
Weed control	2 ridge-up & knock down		@	93 £/ha	186		(75)
Casual labour - planting	30 h/ha	(12 h/ac)	@	14.00 £/h	420		(170)
- harvest & grade	14 t/ha	(6 t/ac)	@	31 £/t	431		(175)
Transport/bulk	13 t/ha	(5.3 t/ac)	@	56 £/t	745		(301)
Other	See notes				50		(20)
Total variable costs						**4254**	(1722)
Gross margin						**3546**	(1435)
Adjustment for wholesale sales							
Marketable yield/output	13 t/ha	(5.3 t/ac)	@	800 £/t	10640		(4306)
less commission @	15%				1596		(646)
Additional casual labour	13 t/ha	1 h/t	@	14.00 £/t	186		(74)
Packaging	13 t/ha	40 bags/t	@	60 p/bag	319		(129)
Additional transport	13 t/ha		@	38 £/t	505		(205)
Gross margin (wholesale)						3779	(1529)

Sensitivity analysis	*Change in value (+/-)*	*Change in gross margin*	*Value range Low*	*High*	*Gross margin range Low*	*High*
Marketable yield	1 t/ha	558 *(223)*	10	20	2430 *(984)*	8008 *(3241)*
Packer price	10 £/t	120 *(48)*	600	700	2946 *(1192)*	4146 *(1678)*
Casual labour	10 h/ha	140 *(56)*	30	100	2997 *(1213)*	3977 *(1610)*

Organic
Growers
Alliance
Join us!
Events • Organic Matters Conference • Organic Growers Gathering • Webinars • Representation • Forum • Organic Grower Magazine • Networking • Podcast • Resources
www.organicgrowersalliance.co.uk

Sell Direct
to Local Homes
& Businesses
Ooooby provides software and support for farms, farm shops and veg box schemes to sell direct online.
ooooby.com
mention this ad for 20% off the set up fee
Ooooby

Carrots and parsnips

Output

Price Higher prices are available for bunching carrots (40-65p/500g bunch). Market for parsnips is limited but improving; processing outlets available for both.

Yield Depends on market, climate and location, as well as spring rainfall, weed pressure, irrigation availability and intensity of production (sensitivity analysis).

Variable costs

Seeds Carrots: graded seed or mini pellets, population 1.5 to 2.5 million/ha. Costs will vary according to variety, growing system and marketing outlet.

Varieties Depends on outlet, early carrots and early maincrop are commonly grown, a vigorous canopy is desirable to suppress weed growth and rapid bulking is needed if late sowing is used to avoid first generation carrot fly attack. Strong top growth is important when using a top harvester.

Irrigation May be needed. Overall cost of applying 25 mm/ha can range from £90-170/ha.

Crop protection Thermal weed control immediately before emergence. Hand weeding will vary according to soil type and weed pressure ranging from 100-500 h/ha and may be higher for early carrots (see Section 12 for costs for weeding equipment).

Harvest Mechanical harvest assumed, for bunching carrots hand harvest at 30-50 h/t. Contract harvesting charges £300-375/ha.

Other Occasional foliar dressings of seaweed, trace elements, pest control. Additional costs for fleece for early carrots would be needed (£1.66/m^2). Additional costs of removing fleece for weeding £90-180/ha.

Crop management

Rotation After cereals or other vegetable crops.

Establishment Conventional primary and secondary cultivations, ridging and de-stoning (if required). Carrots precision-drilled into ridges or a bed system, requiring specialist equipment. Parsnips precision-drilled and rolled.

Nutrition Farm-sourced manure or compost previously in rotation. For carrots, avoid very high fertility and direct manure applications which can increase cavity spot and fanging. Parsnip can also be affected by fanging.

Weed control False seedbed with weed strikes to reduce competition pre-drilling. Thermal weed control immediately prior to emergence (timeliness critical). Brush-weeder or inter-row cultivation; brush, finger-tine weeders or steerage hoes in bed systems. Hand-weeding within rows may be required (1-2 passes) especially in double or triple bands. Keeping totally weed free is not essential; weeding should be timed to coincide with first flush of weed emergence.

Pest control Late sowing to avoid 1st generation carrot fly, or early sowing using fleece/protective netting. Forecasting systems for sowing to avoid carrot fly. Later sowing of parsnips (mid-May) reduces risk of canker. Black surf and powdery scab and viruses can be reduced by using clean seed.

Storage Suitable facilities will be required. In field storage will require straw/plastic or earthing up (costing £2-5000/ha depending on method).

Carrots						£/ha	*(£/ac)*
Marketable yield	25 t/ha	*(10.0 t/ac)*	@	450 £/t	11250		*(4500)*
Total output	75% of gross yield. Gradeout:		10% farm	15% packer		**11250**	*(4500)*
Seed	2 M/ha	*(0.8 M/ac)*	@	1020 £/M	2040		*(816)*
Fertilisers	Standard rotational with additional potash/FYM				82		*(33)*
Flame weeding	1 pass, see notes		@	184 £/ha	184		*(74)*
Brush weeding	2 passes, see notes		@	128 £/ha	256		*(102)*
Casual labour - weeding	250 h/ha	*(100 h/ac)*	@	14.00 £/h	3500		*(1400)*
- harvest & grade	33 t/ha	2 h/t	@	14.00 £/h	231		*(92)*
Transport/bulk	30 t/ha	*(12.0 t/ac)*	@	56 £/t	1680		*(672)*
Other	See notes				50		*(20)*
Total variable costs						**8023**	*(3209)*
Gross margin						**3227**	*(1291)*
Adjustment for wholesale sales							
Marketable yield/output	30 t/ha	*(12.0 t/ac)*	@	650 £/t	19500		*(7800)*
less commission @	15%				2925		*(1170)*
Additional casual labour	30 t/ha	1 h/t	@	14.00 £/h	420		*(168)*
Packaging	30 t/ha	80 bags/t	@	34 p/bag	816		*(326)*
Additional transport	30 t/ha		@	38 £/t	1140		*(456)*
Gross margin (wholesale)						6176	*(2470)*

Sensitivity analysis For explanation, see p. 104

	Change in value (+/-)	*Change in gross margin*	*Value range Low*	*Value range High*	*Gross margin range Low*	*Gross margin range High*
Marketable yield	1 t/ha	345 *(138)*	15	50	-228 *-(91)*	11863 *(4745)*
Packer price	10 £/t	250 *(100)*	300	450	-523 *-(209)*	3227 *(1291)*
Casual labour	10 h/ha	140 *(56)*	150	750	-2849 *-(1140)*	5551 *(2220)*

Parsnips						£/ha	*(£/ac)*
Marketable yield	18 t/ha	*(7.2 t/ac)*	@	700 £/t	12600		*(5040)*
Total output	75% of gross yield. Gradeout:		10% farm	15% packer		**12600**	*(5040)*
Seed	5 kg/ha	*(2 000/ac)*	@	150 £/kg	750		*(300)*
Fertilisers	Standard rotational with additional potash/FYM				82		*(33)*
Flame weeding	1 pass, see notes		@	184 £/ha	184		*(74)*
Brush weeding	1 passes, see notes		@	128 £/ha	128		*(51)*
Casual labour -weeding	250 h/ha	*(100 h/ac)*	@	14.00 £/h	3500		*(1400)*
- harvest/grade	24 t/ha	2 h/t	@	14.00 £/h	672		*(269)*
Transport	20 t/ha		@	56 £/t	1120		*(448)*
Other	See notes				50		*(20)*
Total variable costs						**6486**	*(2594)*
Gross margin						**6114**	*(2446)*
Adjustment for wholesale sales							
Marketable yield/output	20 t/ha	*(8.0 t/ac)*	@	800 £/t	16000		*(6400)*
Less commission @	15 %				2400		*(960)*
Additional casual labour	20 t/ha	1 h/t	@	14.00 £/h	280		*(112)*
Packaging	20 t/ha	80 bags/t	@	34 p/bag	544		*(218)*
Additional transport	20 t/ha		@	38 £/t	760		*(304)*
Gross margin (wholesale)						5530	*(2212)*

Sensitivity analysis For explanation, see p. 104

	Change in value (+/-)	*Change in gross margin*	*Value range Low*	*Value range High*	*Gross margin range Low*	*Gross margin range High*
Marketable yield	1 t/ha	595 *(238)*	10	30	1350 *(540)*	13259 *(5304)*
Packer price	10 £/t	180 *(72)*	650	800	5214 *(2086)*	7914 *(3166)*
Casual labour	10 h/ha	140 *(56)*	150	700	486 *(194)*	8186 *(3274)*

Beetroot and table swedes

Output

Price Beetroot has a limited pre-pack and wholesale market, but there are processing markets suitable for large-scale growers. Price from packer of £300-450/t. Table swedes have reasonable market opportunities in all outlets, but prices can be volatile and change considerably throughout the year.

Yield Depends on market, climate and location, as well as spring rainfall, irrigation availability and intensity of production; see sensitivity analysis for range.

Variable costs

Varieties Beetroot may be round or long rooted depending on market. Long rooted have better storage capability; round ones are better for bunching. Golden and stripy types available.

Weeds Assume flame and brush weeding (see Section 12 for costing).

Harvest Additional costs for mechanical harvest at approximately £40/t.

Other Occasional foliar dressings of seaweed, trace elements, pest control.

Crop management

Rotation After cereals or other vegetable crops. Beetroot can be grown with other roots such as carrots. Swede is a brassica and must be timed to avoid conflict with other brassicas in the rotation.

Establishment Conventional primary and secondary cultivations, ridging and de-stoning (if required). Seeds precision-drilled and rolled. Both crops can also be raised in modules and planted to follow an early crop, but great care is needed to avoid root leader damage.

Nutrition Farm-sourced manure or compost previously in rotation. For beetroots, avoid very high fertility situations and direct manure applications due to nitrate accumulation problems, especially for baby food market.

Weed control False seedbed to reduce weed competition pre-drilling. Brush-weeder or inter-row cultivation/ridging post-emergence if ridge system used. Brush, finger-tine weeder or steerage hoes for inter-row control in bed systems. Hand-weeding and hoeing within rows may be required once or twice depending on season, effectiveness of pre-emergence control and weed pressure. For swedes, inter-row cultivation with harrows or tractor-drawn hoes can be used. The number of passes depends on weed competitiveness but may be 3 or more. Late season weediness is unlikely to affect yields but will make harvesting more difficult.

Pest control Fleece or mesh crop covers may help protect swedes against cabbage root fly and flea beetle. Forecasting systems for cabbage root fly are available from Warwick Crop Centre. Minor root fly damage can be trimmed at harvest. A suitable rotational break is required between swedes and other brassica crops.

Harvest Mechanical, carting, clamping, graded and netted for wholesale or bulk crates for pre-packer.

Storage Suitable facilities will be required. .

Beetroot

					£/ha		*(£/ac)*
Marketable yield	25 t/ha	*(10.0 t/ac)*	@	350 £/t	8750		*(3500)*
Total output	75% of gross yield.	Gradeout:	10% farm		15% processor	**8750**	*(3500)*
Seed	7 kg/ha	*(2.8 kg/ac)*	@	220 £/kg	1540		*(616)*
Fertilisers	Standard rotational with additional potash/FYM				82		*(33)*
Flame weeding	1 pass		@	184 £/ha	184		*(74)*
Brush weeding	2 passes		@	128 £/ha	256		*(102)*
Casual labour -weeding	150 h/ha	*(60 h/ac)*	@	14.00 £/h	2100		*(840)*
- harvest/grade	33 t/ha	2 h/t	@	14.00 £/h	933		*(373)*
Transport	25 t/ha		@	56 £/t	1400		*(560)*
Other	See notes				50		*(20)*
Total variable costs						**6546**	*(2618)*
Gross margin						**2204**	*(882)*
Adjustment for wholesale sales							
Marketable yield/output	25 t/ha	*(10.0 t/ac)*	@	500 £/t	12500		*(5000)*
less commission @	15 %				1875		*(750)*
Additional casual labour	25 t/ha	1 h/t	@	14.00 £/h	350		*(140)*
Packaging	25 t/ha	80 nets/t	@	21 p/net	420		*(168)*
Additional transport	25 t/ha		@	38 £/t	950		*(380)*
Gross margin (wholesale)						2359	*(944)*

Sensitivity analysis For explanation, see p. 104

	Change in value (+/-)	*Change in gross margin*	*Value range Low*	*High*	*Gross margin range Low*	*High*
Marketable yield	1 t/ha	245 *(98)*	10	30	-1478 *-(591)*	3432 *(1373)*
Price	10 £/t	250 *(100)*	300	450	954 *(382)*	4704 *(1882)*
Casual labour	10 h/ha	140 *(56)*	100	400	-362 *-(145)*	3838 *(1535)*

Table swedes

					£/ha		*(£/ac)*
Marketable yield	25 t/ha	*(10.0 t/ac)*	@	380 £/t	9500		*(3800)*
Total output	80% of gross yield. Gradeout:		5% farm		15% packer	**9500**	*(3800)*
Seeds	3.0 kg/ha	*(1.2 kg/ac)*	@	170 £/kg	510		*(204)*
Fertilisers	Standard rotational with additional potash/FYM				82		*(33)*
Brush weeding	1 pass		@	128 £/ha	128		*(51)*
Casual labour -weeding	100 h/ha	*(40 h/ac)*	@	14.00 £/h	1400		*(560)*
- harvest	31 t/ha	2 h/t	@	14.00 £/h	868		*(347)*
Transport/bulk boxes	29 t/ha	*(11.6 t/ac)*	@	56 £/t	1624		*(650)*
Other	See notes				50		*(20)*
Total variable costs						**4662**	*(1865)*
Gross margin						**4838**	*(1935)*
Adjustment for wholesale sales							
Marketable yield/output	29 t/ha	*(11.6 t/ac)*	@	500 £/t	14500		*(5800)*
less commission @	15 %				2175		*(870)*
Additional casual labour	31 t/ha	1 h/t	@	14.00 £/h	434		*(174)*
Packaging	29 t/ha	80 nets/t	@	21 p/net	487		*(195)*
Additional transport	29 t/ha		@	38 £/t	1102		*(441)*
Gross margin (wholesale)						5640	*(2256)*

Sensitivity analysis For explanation, see p. 104

	Change in value (+/-)	*Change in gross margin*	*Value range Low*	*High*	*Gross margin range Low*	*High*
Marketable yield	1 t/ha	279 *(111)*	15	35	2053 *(821)*	7623 *(3049)*
Packer price	10 £/t	250 *(100)*	240	440	1338 *(535)*	6338 *(2535)*
Casual labour	10 h/ha	140 *(56)*	50	400	1506 *(602)*	6406 *(2562)*

Leeks

Output

Price Early crops may attract additional premium, but usually associated with lower yields and higher 'presentation' packaging costs.

Yield Marketable yield will be affected by degree of trimming for specific market outlet. Trimming requirement is influenced by variety; see sensitivity analysis for range.

Variable costs

Seeds/ transplants Assumed modules planted with a row planter, but bare root transplants, also used. 85 plants/m seed row. Cost around £35/1000.

Varieties Different requirements to suit different markets must be considered. Erect varieties allow more frequent mechanical weeding, and some varieties are less susceptible to rust. F1 hybrids have advantages in performance and uniformity.

Irrigation Often needed for establishment. Overall cost of applying 25 mm/ha can range from £90-170/ha.

Weeding Brush weeding assumed but may also involve steerage hoeing (see Section 12). Hand weeding will vary according to soil type and weed pressure ranging from 50-200 h/ha.

Casual labour Harvesting and initial trimming is most commonly done by hand in the field. The smaller the leek the higher the cost. Additional costs incurred for washing, netting, weighing and packing for wholesale sales.

Other Occasional foliar dressings of seaweed and trace elements.

Crop management

Rotation After cereals or other vegetable crops (not onions).

Establish-ment Conventional primary and secondary cultivations. Planting is common, direct drilling may be possible in soils with low weed pressure.

Nutrition Farm-sourced manure or compost previously in rotation.

Weed control False seedbed to reduce weed competition pre-planting. Thermal weed control pre-planting. Brush and finger weeding during early establishment phase. Thermal weed control may be used with care (e.g., shields) post-emergence. Inter-row cultivation with harrows, finger-tine weeders or tractor-drawn hoes – shallow ridging helps to control weeds as well as increasing the length of blanch. The number of passes depends on weed competitiveness. Hand hoeing within rows may be required.

Disease control Maintain appropriate rotational break from other alliums (a minimum of four years from planting to planting). In some years, rust incidence may be high necessitating additional trimming.

Pest control Leek moth and allium leaf miner damage increasing. This can cause economic damage and crop covers may be necessary where there is a risk; thrips can cause a fine white mottling on leaf surfaces which may require closer trimming in bad years.

Harvest Mechanical undercutting and hand lifting and trimming is assumed. On a larger scale mechanical harvesting will reduce costs.

Transport Higher costs for wholesale have been assumed, costs are high because of the difficulty of loading bulk leeks onto pallets.

Leeks					*£/ha*	*(£/ac)*
Marketable yield	14 t/ha	*(5.6 t/ac)*	@ 1500 £/t	21000		*(8400)*
Total output	70% of gross yield. Gradeout:		15% farm	15% packer	21000	*(8400)*
Modules	100 000/ha	*(40 000/ac)*	@ 54 £/000	5400		*(2160)*
Fertilisers	Standard rotational with additional potash/FYM			82		*(33)*
Brush weeding	2 passes		@ 128 £/ha	256		*(102)*
Casual labour - plant	100 h/ha	*(40 h/ac)*	@ 14.00 £/h	1400		*(560)*
- weeding	100 h/ha	*(40 h/ac)*	@ 14.00 £/h	1400		*(560)*
- harvest/trim	20 t/ha	40 h/t	@ 14.00 £/h	11200		*(4480)*
Transport	17 t/ha	*(6.8 t/ac)*	@ 56 £/t	952		*(381)*
Other	See notes			50		*(20)*
Total variable costs					20740	*(8296)*
Gross margin					**260**	*(104)*
Adjustment for wholesale sales						
Marketable yield/output	17 t/ha	*(6.8 t/ac)*	@ 1800 £/t	30600		*(12240)*
less commission @	15 %			4590		*(1836)*
Additional casual labour	17 t/ha	5 h/t	@ 14.00 £/h	1190		*(476)*
Packaging	17 t/ha	80 nets/t	@ 21 p/net	286		*(114)*
Additional transport	17 t/ha		@ 38 £/t	646		*(258)*
Gross margin (wholesale)					3148	*(1259)*

Sensitivity analysis For explanation, see p. 104

	Change in value (+/-)	*Change in gross margin*	*Value range Low*	*Value range High*	*Gross margin range Low*	*Gross margin range High*
Marketable yield	1 t/ha	632 *(253)*	6	18	-4796 -*(1918)*	2788 *(1115)*
Packer price	10 £/t	140 *(56)*	1100	1600	-5340 -*(2136)*	1660 *(664)*
Casual labour	10 h/ha	140 *(56)*	400	1000	260 *(104)*	8660 *(3464)*

Onions

Output

Price	Assumed bulk sales to packer, graded at pack house. Quantity/storage trade-off with price may be possible.
Yield	In bad years, downy mildew can result in substantial crop losses.

Variable costs

Seeds	Assumed using modules, sets are also commonly used (these are more expensive) and direct drilling may be a cheaper alternative, which will have higher weeding costs, but can be effective in a low weed seed soil.
Varieties	A range is available, some with good storage potential. Sets tend to be earlier maturing although storage potential is improving. Module grown onions store better, red varieties are becoming more popular.
Fertilisers	Additional fertilisers as for potatoes.
Irrigation	May be needed; overall cost of applying 25 mm/ha can range from £90-170/ha.
Weeds	Thermal weed control, and brush weeding (see Section 12).
Drying	Costs include loading in and out of store, gas and electricity. £40/t assumed but drying and storage costs may be up to £90/tonne depending on moisture and length of storage.
Harvest	Contract harvesting charges of approximately £40/t would be additional.
Contracting	Contract grading and drying may be a cheaper alternative in some areas.
Other	Occasional foliar dressings of seaweed, trace elements, disease control.

Crop management

Rotation	After cereals, other vegetable crops (not leeks), ley or green manure. Planting after brassicas can suppress onion white rot.
Establish-ment	Conventional primary and secondary cultivations. Module or set planters available. Soil should be firm for sets
Nutrition	Farm-sourced composted manure, Mg and K as indicated by soil analysis.
Weed control	Secondary cultivations for weed control prior to planting. Thermal weed control may be used with care (e.g., shields) post-emergence. Inter-row cultivation with brush weeders, finger-tine weeders, harrows or tractor-drawn hoes. The number of passes depends on weed competitiveness. Hand hoeing may also be required within rows, budgeted at 150h/ha, double in wet years.
Disease control	Maintain appropriate rotational break from other alliums (a minimum of four years between plantings). In some years, neck rot may result in high storage losses. Land known to carry onion white rot should not be used for at least 20 years.
Harvest	Topping, windrowing and mechanical lifting, carting, rimming and netting for wholesale. If onions are grown for storage, they should not be topped to avoid neck rot.
Drying	Dedicated onion dryers, low tech systems (small-scale) and field (skin quality is of paramount importance to successful sales).

Maincrop onions						£/ha		*(£/ac)*
Marketable yield	25 t/ha	*(10.0 t/ac)*	@	530 £/t		13250		*(5300)*
Total output	78% of gross yield.	Gradeout:	0% farm		22% packer		**13250**	*(5300)*
Multiblocks	75 000/ha	*(30 000/ac)*	@	34 £/000		2550		*(1020)*
Fertilisers	Standard rotational with additional potash/FYM					82		*(33)*
Thermal weed control	1 pass flame weeder		@	184 £/ha		184		*(74)*
Brush weeding	1 passes		@	128 £/ha		128		*(51)*
Casual labour - plant	100 h/ha	*(40 h/ac)*	@	14.00 £/h		1400		*(560)*
- weeding	150 h/ha	*(60 h/ac)*	@	14.00 £/h		2100		*(840)*
Drying	32 t/ha	*(12.8 t/ac)*	@	40 £/t		1280		*(512)*
Transport/bulk boxes	32 t/ha	*(12.8 t/ac)*	@	56 £/t		1792		*(717)*
Other	See notes					50		*(20)*
Total variable costs							**9566**	*(3826)*
Gross margin							**3684**	*(1474)*
Adjustment for wholesale sales								
Marketable yield/output	32 t/ha	*(12.8 t/ac)*	@	750 £/t		24000		*(9600)*
less commission @	15%					3600		*(1440)*
Grading	32 t/ha	3.0 h/t	@	14.00 £/h		1344		*(538)*
Packaging (nets)	32 t/ha	40 nets/t	@	23 p/net		294		*(118)*
Additional transport	32 t/ha		@	38 £/t		1216		*(486)*
Gross margin (wholesale)							7979	*(3192)*

Sensitivity analysis	*Change in value (+/-)*	*Change in gross margin*	*Value range Low*	*High*	*Gross margin range Low*	*High*
Marketable yield	1 t/ha	407 *(163)*	20	30	1649 *(660)*	5718 *(2287)*
Packer price	10 £/t	250 *(100)*	450	650	1684 *(674)*	6684 *(2674)*
Casual labour	10 h/ha	140 *(56)*	200	500	184 *(74)*	4384 *(1754)*

Cauliflower and calabrese

Output

Price Highly variable according to season and market outlet. Calabrese subject to oversupply in August, prices for frozen are higher but limited market.

Yield Influenced by market specifications and trimming; see sensitivity analysis.

Variable costs

Seeds Plant populations will vary with the intended market outlet..

Varieties Cauliflower: a large range of types for all seasons.
Calabrese: F1 varieties are standard.

Fertilisers Mineral fertilisers as for potatoes, lime as necessary, may need manganese supplements on soils with very high organic matter.

Irrigation May be needed; Overall cost of applying 25 mm/ha can range from £90-170/ha.

Pest control Assumed use of *Bacillus thuringiensis* (Bt) and potassium soap. Netting for pest control and for early season production is frequently used. Cost £1.66/m^2

Weeds Assumed 3 passes with finger-weeder but may not be required in all years; inter-row weeding may be an alternative. Hours needed for hand weeding vary with date of planting and weather, less in good years.

Harvest Hand harvest assumed; rig harvesting is common for pre-pack crops.

Other Occasional foliar dressings of seaweed.

Crop management

Rotation After cereals, other vegetable crops (not other brassicas), ley or green manure

Establishment Conventional primary and secondary cultivations. Planting with manually operated two or more row planter.

Nutrition Farm-sourced manure or compost

Weed control Secondary cultivations prior to planting (weed strike); inter-row cultivations using brush weeders, finger-tine weeders or tractor-drawn hoes where a slight ridging effect will help bury weeds. Number of passes depends on weed competitiveness, canopy growth and timeliness. Some hand hoeing may be required within rows in early stages.

Disease control Maintain appropriate rotational break from other brassicas (club root risk) and maintain soil pH above 6.5.

Pest control Aphid control through encouragement of natural predators (flowering margins), monitoring/forecasting schemes; natural insecticides only as a last resort. 1-2 applications of potassium soap can sometimes be effective against mealy aphids with good timing. Fleece or mesh crop covers are often used. Can be removed after 2 weeks to allow access for beneficials. *Bacillus thuringiensis* (Bt) as biological control for caterpillars. If birds are a major problem, a range of scarers should be used to get the best effect. Cabbage root fly forecasts available from Warwick Crop Centre; if crop is attacked, extra irrigation and ridging up of soil can limit damage. A green understorey can help to mitigate root fly damage–sowing an appropriate plant in the same module or by undersowing in the crop. These are tricky techniques to get right but it may be worth experimenting with in small areas of crop.

Cauliflower

					£/ha		*(£/ac)*
Marketable yield	1150 doz/ha	*(460 doz/ac)*	@	10.50 £/doz	12075		*(4830)*
Total output	90% of gross yield. (Approx 40% gradeout in field)	Gradeout:		0% farm	10% packer	**12075**	*(4830)*
Transplant modules	30 000/ha	*(12 000/ac)*	@	72 £/000	2160		*(864)*
Fertilisers	Standard rotational with additional potash/FYM				82		*(33)*
Crop protection	2 applications, see notes		@	155 £/ha	310		*(124)*
Finger weeder	3 passes		@	74 £/ha	222		*(89)*
Casual labour - plant	40 h/ha	*(16 h/ac)*	@	14.00 £/h	560		*(224)*
- weeding	40 h/ha	*(16 h/ac)*	@	14.00 £/h	560		*(224)*
- harvest/trim	1278 doz/ha	0.08 h/doz	@	14.00 £/h	1431		*(573)*
Packaging	1278 boxes/ha		@	90 p/box	1150		*(460)*
Transport	1278 doz/ha	*(511 doz/ac)*	@	0.95 £/doz	1208		*(483)*
Other	See notes				50		*(20)*
Total variable costs						**7733**	*(3093)*
Gross margin						**4342**	*(1737)*
Adjustment for wholesale sales							
Marketable yield/output	1278 doz/ha	*(511 t/ac)*	@	12 £/doz	15336		*(6134)*
less commission @	15%				2300		*(920)*
Gross margin (wholesale)						5302	*(2121)*

Sensitivity analysis For explanation, see p. 104

	Change in value (+/-)	*Change in gross margin*	*Value range Low*	*High*	*Gross margin range Low*	*High*
Marketable yield	10 doz/ha	82 *(33)*	1000	1500	3111 *(1244)*	7214 *(2885)*
Price	1 £/doz	1150 *(460)*	9.00	14.00	2617 *(1047)*	8367 *(3347)*
Casual labour	10 h/ha	140 *(56)*	100	350	1993 *(797)*	5493 *(2197)*

Calabrese

					£/ha		*(£/ac)*
Marketable yield	5 t/ha	*(1.9 t/ac)*	@	1500 £/t	7125		*(2850)*
Total output	90% of gross yield.	Gradeout:		0% farm	10% packer	**7125**	*(2850)*
Transplants (modules)	40 000/ha	*(16 000/ac)*	@	47 £/000	1880		*(752)*
Fertilisers	Standard rotational with additional potash/FYM				82		*(33)*
Crop protection	2 applications, see notes		@	172 £/ha	344		*(138)*
Finger weeder	2 passes		@	74 £/ha	148		*(59)*
Casual labour - plant	35 h/ha	*(14 h/ac)*	@	14.00 £/h	490		*(196)*
- weeding	30 h/ha	*(12 h/ac)*	@	14.00 £/h	420		*(168)*
- harvest/grade	5 t/ha	20 h/t	@	14.00 £/h	1400		*(560)*
Transport/bulk boxes	5 t/ha	*(2.0 t/ac)*	@	56 £/t	280		*(112)*
Other	See notes				45		*(18)*
Total variable costs						**5089**	*(2036)*
Gross margin						**2036**	*(814)*
Adjustment for wholesale sales							
Marketable yield/output	5 t/ha	*(2.0 t/ac)*	@	2500 £/t	12500		*(5000)*
less commission @	15%				1875		*(750)*
Packaging	5 t/ha	*(2.0 t/ac)*	@	130 £/t	650		*(260)*
Additional transport	5 t/ha	*(2.0 t/ac)*	@	38 £/t	190		*(76)*
Gross margin (wholesale)						4696	*(1878)*

Sensitivity analysis For explanation, see p. 104

	Change in value (+/-)	*Change in gross margin*	*Value range Low*	*High*	*Gross margin range Low*	*High*
Marketable yield	1 t/ha	1127 *(451)*	3	7	64 *(26)*	4571 *(1828)*
Price	10 £/t	48 *(19)*	1400	2000	1561 *(624)*	4411 *(1764)*
Casual labour	10 h/ha	140 *(56)*	150	400	-1254 -*(502)*	2246 *(898)*

Cabbages

Output

Price Highly variable, depending on outlet but prices of between £550 and £650 per tonne could be considered normal for 2023. See sensitivity analysis for each crop. Savoy cabbage priced per head – often used to offer variety for wholesale.

Yield Influenced by market size and trimming requirements - oversized heads may attract penalty or be rejected. Marketable yield based on crop cut, commonly only 70% of crop grown.

Variable costs

Seeds Plant populations need to be aimed at the intended market.

Varieties A large range of types for all seasons. Mostly F1 hybrids used for uniformity.

Irrigation May be needed; Overall cost of applying 25 mm/ha can range from £90-170/ha.

Pest control Assumed use of *Bacillus thuringiensis* (Bt) and potassium soap. Fleece may be required.

Weeds Assumed 2 passes with finger-weeder. Alternative inter-row weeding may be used. Hours needed for hand weeding will be variable, but many brassicas can be grown without hand weeding in the right conditions.

Other Cold storage costs are expected to increase significantly due to their high energy demands, as well as rising fuel prices. Expected £4.5-£9/t/week for red and white cabbage.

Crop management

Rotation After cereals, other vegetable crops (not other brassicas), ley or green manure

Types Summer (pointed), late summer/autumn (green, white and red), winter (Savoy, January King, Savoy/white crosses such as Tundra).

Establishment Conventional primary and secondary cultivations. Planting with multiple row-planter.

Nutrition Farm-sourced manure or compost

Weed control Secondary cultivations for weed control prior to planting. Subsequently by inter-row cultivations using brush weeders, finger-tine weeders or hoes where a slight ridging effect will help bury weeds. The number of passes depends on weed competitiveness and timeliness.

Disease control Maintain appropriate rotational break from other brassicas (club root risk) and maintain soil pH above 6.5.

Pest control Aphid pest control through flowering field margins and encouragement of natural predators, occasionally potassium soap. Monitoring programmes using traps and/or forecasting schemes can be effective in aiding decisions. Permitted natural insecticides should only be used as a last resort. Bt as biological control for caterpillars should only require one application with good timing and placement. Cabbage root fly forecasts are available from Warwick Crop Centre. Where cabbage root fly attacks crop, extra irrigation and ridging up can limit damage.

Harvest Hand cutting, carting, trimming, netting. Bulk crates may be used for packhouse cabbages, extreme care is required during harvesting to avoid high crop losses in long-term stores.

Storage Long-term cold storage will be required for white and red cabbages

Savoy cabbage

						£/ha	(£/ac)
Marketable yield	28 000/ha	*(11.2 000/ac)*	@	0.6 £/head	16800		*(6720)*
Total output	92% of cut yield.	Gradeout:	0% farm	8% packer		**16800**	*(6720)*
Modules	50 000/ha	*(20 000/ac)*	@	62 £/000	3100		*(1240)*
Fertilisers	Standard rotational with additional potash/FYM				82		*(33)*
Crop protection	2 applications, see notes		@	40 £/ha	80		*(32)*
Finger weeder	3 passes		@	74 £/ha	222		*(89)*
Casual labour - plant	60 h/ha	*(24 h/ac)*	@	14.00 £/h	840		*(336)*
- weeding	40 h/ha	*(16 h/ac)*	@	14.00 £/h	560		*(224)*
- harvest/trim	30 000/ha	7 h/000	@	14.00 £/h	2940		*(1176)*
Transport	30 000/ha	*(12.0 000/ac)*	@	94 £/000	2820		*(1128)*
Packaging	30 000/ha	12 /box	@	0.9 £/box	2250		*(900)*
Other	See notes				50		*(20)*
Total variable costs						**12944**	*(5178)*
Gross margin						**3856**	*(1542)*
Adjustment for wholesale sales							
Marketable yield/output	30 000/ha	*(12.0 000/ac)*	@	0.7 £/head	21000		*(8400)*
Commission	15%				3150		*(1260)*
Gross margin (wholesale)						4906	*(1962)*

Sensitivity analysis For explanation, see p. 104

	Change in value (+/-)	*Change in gross margin*	*Value range Low*	*High*	*Gross margin range Low*	*High*
Marketable yield	1 000/ha	466 *(187)*	20	40	125 *(50)*	9451 *(3781)*
Packer price	0.05 £/head	1400 *(560)*	0.55	0.75	2456 *(982)*	8056 *(3222)*
Casual labour	10 h/ha	140 *(56)*	150	450	1896 *(758)*	6096 *(2438)*

Summer pointed cabbage

						£/ha	(£/ac)
Marketable yield	25 t/ha	*(10.0 t/ac)*	@	580 £/t	14500		*(5800)*
Total output	83% of cut yield.	Gradeout:	5% farm	12% packer		**14500**	*(5800)*
Transplants	70 000/ha	*(28 000/ac)*	@	62 £/000	4340		*(1736)*
Fertilisers	Standard rotational with additional potash/FYM				82		*(33)*
Crop protection	2 applications, see notes		@	40 £/ha	80		*(32)*
Finger weeder	3 passes		@	74 £/ha	222		*(89)*
Casual labour - plant	80 h/ha	*(32 h/ac)*	@	14.00 £/h	1120		*(448)*
- weeding	40 h/ha	*(16 h/ac)*	@	14.00 £/h	560		*(224)*
- harvest/trim	30 t/ha	8 h/t	@	14.00 £/h	3360		*(1344)*
Transport	29 t/ha	*(11.6 t/ac)*	@	56 £/t	1624		*(650)*
Other	See notes				50		*(20)*
Total variable costs						**11438**	*(4575)*
Gross margin						**3062**	*(1225)*
Adjustment for wholesale sales							
Marketable yield/output	29 t/ha	*(11.6 t/ac)*	@	650 £/t	18850		*(7540)*
less commission @	15%				2828		*(1131)*
Packaging	29 t/ha	12.5 kg/net	@	21 p/net	487		*(195)*
Additional transport	29 t/ha	*(11.6 t/ac)*	@	38 £/t	1102		*(441)*
Gross margin (wholesale)						2995	*(1198)*

Sensitivity analysis For explanation, see p. 104

	Change in value (+/-)	*Change in gross margin*	*Value range Low*	*High*	*Gross margin range Low*	*High*
Marketable yield	1 t/ha	381 *(152)*	20	35	1157 *(463)*	6871 *(2749)*
Packer price	10 £/t	250 *(100)*	490	650	812 *(325)*	4812 *(1925)*
Casual labour	10 h/ha	140 *(56)*	200	500	1102 *(441)*	5302 *(2121)*

Red cabbage					£/ha		*(£/ac)*
Marketable yield	25 t/ha	*(10.0 t/ac)*	@	540 £/t	13500		*(5400)*
Total output	75% of cut yield.	Gradeout:	10% farm	15% packer		**13500**	*(5400)*
Transplants	50 000/ha	*(20 000/ac)*	@	55 £/000	2750		*(1100)*
Fertilisers	Standard rotational with additional potash/FYM				82		*(33)*
Crop protection	2 applications, see notes		@	40 £/ha	80		*(32)*
Finger weeding	3 passes		@	74 £/ha	222		*(89)*
Casual labour - plant	65 h/ha	*(26 h/ac)*	@	14.00 £/h	910		*(364)*
- weeding	40 h/ha	*(16 h/ac)*	@	14.00 £/h	560		*(224)*
- harvest/grade	33 t/ha	8 h/t	@	14.00 £/h	3696		*(1478)*
Transport	33 t/ha	*(13.2 t/ac)*	@	56 £/t	1848		*(739)*
Other	See notes				285		*(114)*
Total variable costs						**10433**	*(4173)*
Gross margin						**3067**	*(1227)*
Adjustment for wholesale sales							
Marketable yield/output	33 t/ha	*(13.2 t/ac)*	@	650 £/t	21450		*(8580)*
less commission @	15%				3218		*(1287)*
Packaging	33 t/ha	12.5 kg/net	@	21 p/net	554		*(222)*
Additional transport	33 t/ha	*(13.2 t/ac)*	@	38 £/t	1254		*(502)*
Gross margin (wholesale)						5991	*(2396)*

Sensitivity analysis For explanation, see p. 104

	Change in value (+/-)	*Change in gross margin*	*Value range Low*	*High*	*Gross margin range Low*	*High*
Marketable yield	1 t/ha	755 *(302)*	15	40	-4487 *-(1795)*	14397 *(5759)*
Price	10 £/t	250 *(100)*	395	650	-558 *-(223)*	5817 *(2327)*
Casual labour	10 h/ha	140 *(56)*	200	500	1233 *(493)*	5433 *(2173)*

White cabbage					£/ha		*(£/ac)*
Marketable yield	35 t/ha	*(14.0 t/ac)*	@	540 £/t	18900		*(7560)*
Total output	75% of gross yield.	Gradeout:	10% farm	15% packer		**18900**	*(7560)*
Transplants	60 000/ha	*(24 000/ac)*	@	62 £/000	3720		*(1488)*
Fertilisers	Standard rotational with additional potash/FYM				82		*(33)*
Crop protection	2 applications, see notes		@	40 £/ha	80		*(32)*
Finger weeder	3 passes		@	74 £/ha	222		*(89)*
Casual labour - plant	70 h/ha	*(28 h/ac)*	@	14.00 £/h	980		*(392)*
- weeding	40 h/ha	*(16 h/ac)*	@	14.00 £/h	560		*(224)*
- harvest/grade	47 t/ha	8 h/t	@	14.00 £/h	5264		*(2106)*
Transport/bulk boxes	42 t/ha	*(16.8 t/ac)*	@	56 £/t	2352		*(941)*
Other	See notes				285		*(114)*
Total variable costs						**13545**	*(5418)*
Gross margin						**5355**	*(2142)*
Adjustment for wholesale sales							
Marketable yield/output	42 t/ha	*(16.8 t/ac)*	@	650 £/t	27300		*(10920)*
less commission @	15%				4095		*(1638)*
Packaging	42 t/ha	18 kg/net	@	21 p/net	490		*(196)*
Additional transport	42 t/ha		@	38 £/t	1596		*(638)*
Gross margin (wholesale)						7574	*(3029)*

Sensitivity analysis For explanation, see p. 104

	Change in value (+/-)	*Change in gross margin*	*Value range Low*	*High*	*Gross margin range Low*	*High*
Marketable yield	1 t/ha	323 *(129)*	20	50	503 *(201)*	10207 *(4083)*
Packer price	10 £/t	350 *(140)*	395	650	280 *(112)*	9205 *(3682)*
Casual labour	10 h/ha	140 *(56)*	250	600	3759 *(1503)*	8659 *(3463)*

Lettuce

Output

Price Highly variable, depending on type (Gem, Cos, Batavia, Iceberg, etc) and time of season. Continuity of supply is essential to ensure stable markets for both wholesale and multiple outlets. This will require pre-programming into market outlet – prior to production

Yield Marketable yield will be affected by degree of trimming for specific markets to remove aphid/slug damage, tip burn and mildew, approx. 30% gradeout in the field. Yield will be lower for second crop where double cropped. Head weights may be more difficult to achieve for Iceberg.

Variable costs

Transplants 60,000-150,000 plants required per hectare.

Planting 2-row planter assumed.

Varieties Mildew resistant varieties available. A wide range of varieties of different types available.

Irrigation Considered essential for programmed production (10-15 mm every 4 days), overall cost of applying 25 mm/ha can range from £90-170/ha.

Pest control Assumed use of potassium soap.

Packaging May also be done on-farm for pre-pack outlets (see wholesale for costs).

Other Expected £5.5-10/t/week (not included), vacuum cooler required for large volumes of Iceberg.

Crop management

Rotation Well placed after manured vegetable crops or green manure. Often double-cropped.

Establishment Conventional primary and secondary cultivations. Planting using self-propelled or tractor-mounted equipment.

Nutrition Potash hungry. Over-wintered green manure top-dressed with 25t FYM/ha (10t/ac). Green waste compost can be considered. An additional nitrogen source may be needed on light soils.

Weed control Front-mounted steerage hoes or brush weeders (typically 2 passes, depending on weed competition and soil type) and hand labour (normally not necessary).

Disease control Maintain an appropriate rotational break of at least 48 months between plantings on the same land as very susceptible to soil-borne diseases.

Pest control Aphid pest control through flowering field margins and encouragement of natural predators, (ladybirds, hoverflies and lacewings), occasionally potassium soap and garlic sprays may be helpful. Slugs and leatherjackets could be a problem in early crops straight after grass.

Harvest Hand harvest, trimming and packing are required. The use of tractor-mounted packing rigs may improve harvesting efficiency. Cold storage is essential. The 'Iceberg' crop must be cooled immediately after harvest if it is to be sold as 'Iceberg' for retailing, otherwise it should be marketed as a bagged crisp lettuce.

Other Fleece cover maybe be required for early crop production.

Lettuce						£/ha	*(£/ac)*
Marketable yield	3350 doz/ha	*(1340 t/ac)*	@	7.00 £/doz	23450		*(9380)*
Total output	88% of gross yield.	Gradeout:		0% farm	12% packer	**23450**	*(9380)*
Transplants (blocks)	80 000/ha	*(32 000/ac)*	@	57 £/000	4560		*(1824)*
Irrigation	75 mm		@	4.8 £/mm	360		*(144)*
Fertilisers	Standard rotational with additional potash/FYM				82		*(33)*
Crop protection	1 applications, see notes		@	45 £/ha	45		*(18)*
Brush weeding	2 pass		@	128 £/ha	256		*(102)*
Casual labour - plant	40 h/ha	*(16 h/ac)*	@	14.00 £/h	560		*(224)*
- weeding	60 h/ha	*(24 h/ac)*	@	14.00 £/h	840		*(336)*
- harvest/trim	3807 doz/ha	0.05 h/doz	@	14.00 £/h	2665		*(1066)*
Transport/bulk boxes	3807 doz/ha	*(1523 doz/ac)*	@	0.95 £/doz	3617		*(1447)*
Commision/other					50		*(20)*
Total variable costs						**13035**	*(5214)*
Gross margin						**10415**	*(4166)*
Adjustment for wholesale sales							
Marketable yield/output	3807 doz/ha	*(1523 doz/ac)*	@	8.50 £/doz	32360		*(12944)*
less commission @	15%				4854		*(1942)*
Packaging	3807 doz/ha	1 doz/bo›	@	90 p/box	3426		*(1371)*
Additional transport	3807 doz/ha	*(1523 doz/ac)*	@	0.95 £/doz	3617		*(1447)*
Gross margin (wholesale)						7428	*(2971)*

Sensitivity Analysis

	Change in value (+/-)	*Change in gross margin*	*Value range Low*	*High*	*Gross margin range Low*	*High*
Marketable yield	100 doz/ha	513 *(205)*	1750	4000	2215 *(886)*	13746 *(5499)*
Price	0.1 £/doz.	335 *(134)*	6.00	10.00	7065 *(2826)*	20465 *(8186)*
Casual labour	10 h/ha	140 *(56)*	200	600	6080 *(2432)*	11680 *(4672)*

Courgettes	*notes on next page*					£/ha	*(£/ac)*
Marketable yield	15.0 t/ha	*(6.0 t/ac)*	@	1100 £/t	16500		*(6600)*
Total output	87% of gross yield. Gradeout:			5% farm	8% packer	**16500**	*(6600)*
Plants	10 000/ha	*(4 000/ac)*	@	56.5 £/000	565		*(226)*
Fertilisers	Standard rotational with additional potash/FYM				82		*(33)*
Brush weeding	2 pass		@	128 £/pass	256		*(102)*
Casual labour - plant	70 h/ha	*(28 h/ac)*	@	14.00 £/h	980		*(392)*
- weeding	50 h/ha	*(20 h/ac)*	@	14.00 £/h	700		*(280)*
- harvest	16.2 t/ha	50 h/t	@	14.00 £/h	11340		*(4536)*
Transport	16.2 t/ha	*(6.5 t/ac)*	@	56 £/t	907		*(363)*
Other	See notes				25		*(10)*
Total variable costs						**14855**	*(5942)*
Gross margin						**1645**	*(658)*
Adjustment for wholesale sales							
Marketable yield/output	16.2 t/ha	*(6 t/ac)*	@	1400 £/t	22680		*(9072)*
Commission	15%				3402		*(1361)*
Packaging	16.2 t/ha	1 doz/box	@	90 p/box	15		*(6)*
Additional transport	16.2 t/ha			38 £/t	615.6		
Gross margin (wholesale)						3792	*(1517)*

Sensitivity analysis For explanation, see p. 104

	Change in value (+/-)	*Change in gross margin*	*Value range Low*	*High*	*Gross margin range Low*	*High*
Marketable yield	1 t/ha	234 *(94)*	8	17	5 *(2)*	2113 *(845)*
Price	10 £/t	150 *(60)*	850	1200	-2105 *-(842)*	3145 *(1258)*
Casual labour	10 h/ha	140 *(56)*	500	800	3465 *(1386)*	7665 *(3066)*

Courgettes

Output

Price Highly variable, higher prices early and late in the season.

Yield Influenced by market specifications and frequency of picking. Gluts can occur in the peak of the season.

Variable costs

Seeds Transplants are commonly used. Drilling direct possible, increased weeding costs likely.

Varieties A number of varieties including F1 hybrids are available as organic seed for local and wholesale markets; F1 are industry standard.

Fertiliser Rotational application, costs for other minerals (as indicated by soil analysis) and additional manure.

Irrigation Necessary for establishment and yield. Overall cost of applying 25mm/ha can range from £90-170/ha.

Other Occasional foliar dressing of seaweed.

Crop management

Rotation After cereals, other vegetable crops, ley or green manure.

Establishment Conventional primary and secondary cultivations. Mechanical planting or direct drilling, or manual planting through mechanically laid black plastic or woven polypropylene groundcover. Fleece used for frost protection for early production

Nutrition Farm sourced manures or compost. Lime if needed well in advance as not tolerant of acidity.

Weed control Secondary cultivation for weed control prior to drilling/planting. Subsequently by inter-row cultivations using brush weeders, spring-tine weeders or tractor-drawn hoes. Hand hoeing may be required within rows. Alternatively, use black plastic mulch or woven polypropylene groundcover with clover, biodegradable mulches, mowing or inter-row cultivation between beds.

Disease control Crop rotation. Powdery mildew is biggest problem reducing yield late in the season. Potassium bicarbonate is likely to be more effective than sulphur. Irrigation during dry spells can also be helpful as water stress increases susceptibility. Space plants to allow good air movement

Pest control Slugs may be a problem under mulches in heavy soils. Few other pest problems.

Harvest Regular harvesting (every day in the height of season) essential even when the market demand is low, requiring considerable commitment of labour.

Other Cold storage is likely to be necessary between harvest and sale. Cold storage costs not included (see lettuce for costs).

Celery

Output

Price Highly variable; continuity of supply is essential to ensure stable markets for both wholesale and multiple outlets. This requires pre-programming into market outlets prior to production.

Yield Marketable yield is affected by percentage cut.

Variable costs

Transplants 50,000-70,000 plants required per hectare

Planting 2-row planter assumed.

Varieties F1s are the industry standard.

Irrigation Considered essential for production, see potatoes for costs.

Other Cold storage (for costs see lettuce).

Crop management

Rotation Well placed after manured vegetable crops or fertility-building.

Establishment Conventional primary and secondary cultivations. Planting using self-propelled or tractor-mounted equipment.

Nutrition Potash hungry. Over-wintered green manure top-dressed with 25t FYM/ha (10t/ac). Rotational fertiliser costs not attributed as double cropped.

Weed control Inter-row hoes or brush weeders (typically 2 passes, depending on weed competition and soil type) and hand labour (as necessary).

Disease control Celery leaf spot (*Septoria*) can be devastating at the end of the season. Healthy seeds and transplants essential. Isolate successive plantings wherever possible.

Pest control Aphid pest control through flowering field margins and encouragement of natural predators (ladybirds hoverflies and lacewings), occasionally potassium soap. Garlic sprays may be helpful. Leaf miner can occasionally be a problem although minor damage will not affect yield and is trimmed at harvest.

Harvest Hand harvest, trimming and packing are required. The use of tractor-mounted packing rigs may improve harvesting efficiency. Cold storage is essential.

Other Fleece cover required for early crop production.

Celery

						£/ha	*(£/ac)*
Marketable yield	2750 doz/ha	*(1100 dz/ac)*	@	9.00 £/doz	24750		*(9900)*
Total output	75% of gross yield. Gradeout:		20% farm		5% packer	**24750**	*(9900)*
Modules	60 000/ha	*(24 000/ac)*	@	54 £/000	3240		*(1296)*
Fertilisers	Standard rotational with additional potash/FYM				82		*(33)*
Irrigation	75 mm		@	4.8 £/mm	360		*(144)*
Finger weed	3 passes			74 £/ha	50		*(20)*
Casual labour - plant	70 h/ha	*(28 h/ac)*	@	14.00 £/h	980		*(392)*
- weeding	70 h/ha	*(28 h/ac)*	@	14.00 £/h	980		*(392)*
- harvest and pack	3667 doz/ha	15 doz/h	@	14.00 £/h	3422		*(1369)*
Transport	2895 doz/ha		@	1.26 £/doz	3647		*(1459)*
Packaging	2895 doz/ha		@	1.2 £/doz	3474		*(1389)*
Total variable costs						**16235**	*(6494)*
Gross margin						**8515**	*(3406)*
Adjustment for wholesale sales							
Marketable yield/output	2895 doz/ha	*(1158 t/ac)*	@	9.50 £/doz	27500		*(11000)*
Commission	15%				4125		*(1650)*
Gross margin (wholesale)						7140	*(2856)*

Sensitivity analysis For explanation, see p. 104

	Change in value (+/-)	*Change in gross margin*	*Value range Low*	*High*	*Gross margin range Low*	*High*
Marketable yield	10 doz/ha	71 *(29)*	2500	3500	6729 *(2692)*	13872 *(5549)*
Price	1 £/doz	2750 *(1100)*	7.00	10.40	3015 *(1206)*	12365 *(4946)*
Casual labour	10 h/ha	140 *(56)*	50	400	4875 *(3910)*	9775 *(1950)*

Sweetcorn

Output

Price	Will vary through the season, with higher prices for early production.
Yield	Marketable yield (cobs/ha) will vary according to season and nitrogen availability, see sensitivity analysis for range. Only 75% likely to be cut in the field.

Variable costs

Seeds and planting	Assumed direct drilling, although transplants are also commonly used. These will be more expensive (£52-140/1,000 depending on quantity ordered) but may provide an earlier crop and better weed control, although they are sensitive to late frosts.
Varieties	Supersweet varieties normally grown
Fertilisers	Rotational amounts as per other crops, will benefit from applications of FYM.
Weed control	Assumed 2x inter-row hoeing, can require some hand weeding.
Irrigation	Not essential but can increase yields when applied during silking and when ears are filling. .

Crop management

Rotation	Well placed after manured vegetable crops or fertility-building.
Establish-ment	Conventional primary and secondary cultivations. Usually direct-drilled and rolled but can be transplanted to advance season or in marginal locations
Nutrition	High N requirement. Needs moderate to high levels of phosphorus and potassium. Over-wintered green manure top-dressed with 25t FYM/ha (10t/ac). Rotational fertiliser costs not attributed as double cropped.
Weed control	Stale seedbeds and thermal weed control pre-emergence. Thermal weed control may be used with care (e.g., shields) post-emergence. Brush and/or finger weeding during early establishment phase. Under-sowing with clover is a useful option and can be effective in establishing a fertility break.
Disease control	Unlikely to be necessary.
Pest control	Bird scarers may be needed. Leatherjackets may be problematic following grass. Badgers can be a serious problem, electric netting can be used and may be essential.
Harvest	Hand harvest, trimming and packing are required. The use of tractor-mounted packing rigs may improve harvesting efficiency. Short-term cold storage is useful; for costs see lettuce.
Other	Fleece cover required for early crop production.

Sweetcorn

						£/ha	(£/ac)
Marketable yield	30 000/ha	*(12.0 000/ac)*	@	0.35 £/cob	10500		*(4200)*
Total output	85% of cut yield.	Gradeout:	5% farm	10% packer		**10500**	*(4200)*
Seeds	10 kg/ha	*(4 kg/ac)*	@	32.5 £/kg	325		*(130)*
Fertilisers	Standard rotational with additional potash/FYM				82		*(33)*
Inter row hoe	2 passes			45 £/ha	90		*(36)*
Casual labour - weeding	40 h/ha	*(16 h/ac)*	@	14.00 £/h	560		*(224)*
- harvest	33 000/ha	300 cobs/h	@	14.00 £/h	1540		*(616)*
Transport	33 000/ha		@	37 £/000	1221		*(488)*
Packaging	30 000/ha	30 /box	@	0.9 £/box	900		*(360)*
Other	See notes				50		*(20)*
Total variable costs						**4768**	*(1907)*
Gross margin						**5732**	*(2293)*
Adjustment for wholesale sales							
Marketable yield/output	33 000/ha	*(13.2 000/ac)*	@	0.5 £/cob	16500		*(6600)*
Commission	15%				2475		*(990)*
Additional packaging	33 000/ha	30.0 cobs/box	@	90 p/box	990		*(396)*
Gross margin (wholesale)						8267	*(3307)*

Sensitivity analysis For explanation, see p. 104

	Change in value (+/-)	*Change in gross margin*	*Value range Low*	*High*	*Gross margin range Low*	*High*
Marketable yield	5 000/ha	1125 *(450)*	10	38	1233 *(493)*	7532 *(3013)*
Packer price	0.05 £/head	1500 *(600)*	0.30	0.40	4232 *(1693)*	7232 *(2893)*
Casual labour	10 h/ha	140 *(56)*	50	200	5032 *(2853)*	7132 *(2013)*

Strawberries

Output

Price Prices will vary with crop quality (Class 1 assumed), time of year and weather. An average price used, is based on 70% of Class 1 sales to packers and 30% grade out for processing. The Gross Margin is calculated on a per hectare basis, sold wholesale. Although the area grown is likely to be significantly smaller and must depend on securing the market first.

Yield Estimates are based on growing in vented polytunnels. Lower yield in the maiden year is assumed, average for 2 years is used, but there is potential for a third year of cropping with lower yield and quality. Potential for pick-your-own in the third year with lower quality and yield.

Variable costs

Polytunnels Mobile fully ventilated tunnels. Cost from £20/m². Cost can be depreciated over 10 years. There is now little outdoor production of strawberries.

Establishment Conventional plants are assumed as organic plants are currently unavailable. Derogation is required and may not be granted, in which case home propagation required costing approx. £220/1000. Planted into raised beds (1.5m width) with plastic mulch, and laying of trickle irrigation (T-tape, 1 or 2 lines per bed), run beneath the plastic mulch adjacent to each row. Use of PHPS certified planting material is advisable. Costs vary depending on plant density, mulch material and the irrigation system.

Irrigation Assumed use of trickle irrigation. Equipment and water costs not included..

Harvest Hand picking on the farm into punnets and boxes is assumed.

Other Scales, trolleys, toilet facilities for harvesters etc.

Crop management

Rotations After grass/clover leys (1-2 yrs old), brassicas (not rape), rye, winter cereals (not oats), livestock.

Cultivations Standard primary and secondary cultivations.

Nutrition Farmyard manure or compost applied 25t/ha (10t/ac) at establishment. Mineral fertilisers applied on rotational basis as needed.

Planting 40-45 cm apart (for medium-sized variety) on 2-row bed. Wider plant spacing is desirable for good ventilation in crop canopy and effective disease control. Hand planting assumed.

Weed control Using plastic mulches or woven polypropylene groundcover assumed. 40-45 cm apart (for medium sized variety) on 2-row bed.

Pest control Biological controls for red spider mite and other pests. Very few effective control agents are available. Establishment of wildflower strips and beetle banks to encourage beneficial insects is advised to help control insect pests.

Disease control Plants suffering from more serious diseases such as crown rot should be removed and destroyed. Potassium bicarbonate useful for the control of powdery mildew, likely to be a problem in certain varieties.

Harvest Access to cold store will be required for bulk sales; freezing facilities may be required for processing.

Strawberries

						£/ha	(£/ac)
Maiden year	12500 kg/ha	*(5000 kg/ac)*	@	7.00 £/kg	87500		*(35000)*
First full crop	16600 kg/ha	*(6640 kg/ac)*	@	7.00 £/kg	116200		*(46480)*
Total output	**(2 year av.)**					**101850**	*(40740)*
Establishment (see below)	8744 £/ha	*(3498 £/ac)*	over	2 years	4372		*(1749)*
Mypex	5000 m		@	0.5 £/m	1250		
Labour (maintenance)	420 hr/yr	*(168 h/ac)*	@	14.00 £/h	5880		*(2352)*
Harvest (piecework)	14500 kg/ha	(5 kg/hr	@	14.00 £/hr	40600		*(16240)*
Punnets (250g)			@	0.32 £/kg	4640		*(1856)*
Other	See notes				250		*(100)*
Variable costs	**(2 year annual average)**					**56992**	*(22797)*
Maiden year	Variable costs	44872 £/ha	*(17949 £/ac)*		42628		*(17051)*
First full crop	Variable costs	63544 £/ha	*(25418 £/ac)*		52656		*(21062)*
Gross margin	**(2 year annual average)**					**21314**	*(8526)*
Establishment costs							
Bed raising					350		*(140)*
Woodchip + application labour					1930		*(772)*
Straw mulch + labour	350 bales				1260		*(504)*
Biological pest control	See notes				250		
Fertiliser	Standard rotational with additional potash/FMY				82		*(33)*
Plants	41.6 000/ha	*(17 000/ac)*	@	140 £/1000	2912		*(1165)*
Planting labour	280 hr/ha		@	14.00 £/hr	1960		*(784)*
Establishment costs						**8744**	*(3498)*

Sensitivity analysis For explanation, see p. 104

	Change in value (+/-)	*Change in av. gross margin*	*Value range Low*	*High*	*Average gross margin range Low*	*High*
Marketable yield	100 kg/ha	430 *(172)*	13000	16000	14649 *(5860)*	27764 *(11106)*
Price	0.1 £/kg	1450 *(580)*	5.50	8.00	-436 -*(174)*	35814 *(14326)*
Harvest labour	10 p/kg	1450 *(580)*	40	90	10294 *(4118)*	17544 *(7018)*

Apples and pears[11]

Output

Price	Sale to wholesaler. Packing assumed, attracting highest price but also costs for grading, packing and transport. Price varies with market, variety, quality, time of year and weather. Due to higher costs of production, both apple and pear profitability is sensitive to low prices and this is reflected in the gross margin.
Yield	Average marketable yield or quantity selling for the dessert market in Class I and II (combined class) over the productive life of an orchard. Yields vary with planting density (750-2,500/ha), variety, rootstock, soil type, weather, and age of tree.
Gradeout	Percentage varies with variety and season (range from 50-80%). Wholesale markets will have less strict specifications on size and blemishes. Gradeouts can sell for processing, mainly juice.

Variable costs

Establishment	Based on organic trees, spacing of apples: 3.5m x 2m and pears 3.5m x 2.5m. Costs will vary with planting density. Options include establishing a new organic orchard from scratch or converting an existing one. New disease resistant varieties are available.
Nutrition	Soil fertility maintained as necessary with foliar feeds (e.g., seaweed extract), permitted organic fertilisers (e.g., pelleted chicken manure), mineral fertilisers and lime, mulches of compost (including green waste compost) or farmyard manure. Level of use and costs vary widely between growers. Grass/clover mix for alleyway assumed, but re-establishment costs every 5 years are not included.
Planting	Organically-raised trees are available, but numbers and choice of variety may be limited (some imports are used although transporting live trees is time sensitive and post-EU-exit importation comes with risks). Tree and row spacing will depend on rootstock used and orchard design.
Weed control	Options include cultivated strips and mulches beneath tree rows, or grass sward/ley cover up to trees, mown regularly. Costs not included
Pest and disease control	Establishing wildflower strips/grassy banks to encourage predators and growing resistant varieties is advisable. Pruning to maintain an open tree canopy and good air circulation will help reduce incidence of disease. Use as necessary of sulphur, potassium bicarbonate, pyrethrum and Bt. Level of use and costs vary widely between growers. Certifying body should be consulted prior to application of plant protection products.
Irrigation	Not essential but can increase yields when applied during silking and when ears are filling.
Pruning	Will vary with age and size of trees from 25-150h/ha.
Harvest	Hand picking on the farm into bins.
Grading and packing	Assumed to be done on farm into large boxes.
Other	Includes industry levies.

Apples								£/ha	*(£/ac)*
Prices	70%	Cl. I/II	1 £/kg	30%	Process	0.2 £/kg			
Establishment, yrs 1-5	3	t/ha	*(1.2 t/ac)*	@	760	£/t	2280		
Peak production yrs 6-11	16	t/ha	*(6.4 t/ac)*	@	760	£/t	12160		
Further prodn. yrs 12-20	14	t/ha	*(5.6 t/ac)*	@	760	£/t	10640		
Total output	**11.9**	**t/ha**	**(20 year av.)**					**9006**	*(3602)*
Establishment (see below)	35961	£/ha	*(14384 t/ac)*	over	20	years	1798		*(719)*
Crop protection	see	notes					1000		*(400)*
Fertilisers	see	notes					150		*(60)*
Pruning costs	1428	trees/ha		@	0.8	£/tree	1142		*(457)*
Picking costs	12	t/ha	*(5 t/ac)*	@	85	£/t	1007		*(403)*
Grade and pack	8	t/ha	*(3 t/ac)*	@	175	£/t	1452		*(581)*
Storage and bin hire	37	bins(320kg)		@	34	£/bin	1259		*(504)*
Market commission				@	10	%	901		*(360)*
Transport	12	t/ha	*(5 t/ac)*	@	110	£/t	1304		*(521)*
Other							250		*(100)*
Variable costs	**(20 year annual average)**							**10262**	*(4105)*
Establishment, yrs 1-5	Variable costs	2,598 £/ha	*(1039 £/ac)*				-318		-*(127)*
Peak production yrs 6-11	Variable costs	13,857 £/ha	*(5543 £/ac)*				-1697		-*(679)*
Further prodn. yrs 12-20	Variable costs	12,124 £/ha	*(4850 £/ac)*				-1484		-*(594)*
Gross margin	**(20 year annual average)**							**-1256**	-*(503)*
Adjustment for wholesale sales and other marketing costs									
Prices	75%	Cl. II	1.7 £/kg	25%	Process	0.3 £/kg			
Marketable yield	11.9	t/ha	*(5 t/ac)*	@	1350	£/t	15998		*(6399)*
Additional commission	7.5%						1200		*(480)*
Packaging	11.9	t/ha		@	65	£/t	770		*(308)*
Gross margin (wholesale)								14027	*(5611)*
Establishment costs									
Trees (organic)	1428	/ha	*(571 /ac)*	@	19	£/tree	27132		*(10853)*
Stakes	1428	/ha	*(571 /ac)*	@	1.2	£/tree	1713.6		*(685)*
Guards and ties	1428	/ha	*(571 /ac)*	@	0.85	£/tree	1213.8		*(486)*
Labour to plant	1428	/ha	*(571 /ac)*	@	2.2	£/tree	3141.6		*(1257)*
Drill and establish swards	1	ha		@	465	£/ha	465		*(186)*
Irrigation drip line	2700	m/ha	*(1080 m/ac)*	@	85	p/m	2295		*(918)*
Establishment costs								**35961**	*(14384)*

Sensitivity analysis

	Change in value (+/-)	Change in av. gross margin	Value range Low	Value range High	Average gross margin range Low		High	
Total yield	1 t/ha	543 *(217)*	2	26	-6606	-*(2642)*	6428	*(2571)*
Price	0.1 £/kg	1185 *(474)*	0.9	1.4	403	*(161)*	6328	*(2531)*
Gradeout*	1%	95 *(38)*	40%	80%	-4100	-*(1640)*	-308	-*(123)*

*In fruit growing this represents the quantity reaching Class I & II standard

11. Firth C (2005) Economics of Organic Top Fruit Production. http://orgprints.org/10775/
Organic Fruit Growing (Lind et al 2003). CABI Publishing: Wallingford.
Organic Fruit Production and Viticulture (Cubison, S. 2009), Crowood Press

Pears						£/ha		*(£/ac)*
Prices	70% Cl. I/II	1.2 £/kg		30% Process	0.3 £/kg			
Establishment years 1-6	2 t/ha	*(1 t/ac)*	@	930 £/t		1860		*(744)*
Peak production yrs 7-12	11 t/ha	*(4 t/ac)*	@	930 £/t		10230		*(4092)*
Further prodn. yrs 13-30	9 t/ha	*(4 t/ac)*	@	930 £/t		8370		*(3348)*
Total output	**8 t/ha**	**(30 year av.)**					**7440**	*(2976)*
Establishment (see below)	30478 £/ha	*(12191 £/ac)*	over	30 years		1016		*(406)*
Crop protection	see notes					200		*(80)*
Fertilisers	see notes					125		*(50)*
Pruning costs	1143 trees/ha		@	0.8 £/tree		914		*(366)*
Picking costs	8 t/ha	*(3 t/ac)*	@	85 £/t		680		*(272)*
Grade and pack	6 t/ha	*(2 t/ac)*	@	175 £/t		980		*(392)*
Storage and bin hire	17.5 bins(320kg)		@	34 £/bin		595		*(238)*
Market commission	7440 £/output		@	6 %		446		*(179)*
Transport	8 t/ha	*(3 t/ac)*	@	110 £/t		880		*(352)*
Other						250		*(100)*
Variable costs	**(30 year annual average)**						**6087**	*(2435)*
Establishment years 1-5	Variable costs	1522 £/ha		*(609 £/ac)*		338		*(135)*
Peak production yrs 7-12	Variable costs	8369 £/ha		*(3348 £/ac)*		1861		*(744)*
Further prodn. yrs 13-30	Variable costs	6848 £/ha		*(2739 £/ac)*		1522		*(609)*
Gross margin	**(30 year annual average)**						**1353**	*(541)*
Adjustment for wholesale sales								
Prices	85% Cl. I/II	1.9 £/kg		15% Process	0.3 £/kg			
Marketable yield	8 t/ha	*(3 t/ac)*	@	1660 £/t		13280		*(5312)*
Packaging costs	8 t/ha		@	55 £/t		440		*(176)*
Additional commission	7.5%					33		*(13)*
Gross margin (wholesale)							6753	*(2701)*
Establishment costs								
Trees (organic)	1143 ha	*(457 /ac)*	@	20 £/tree		22860		*(9144)*
Stakes	1143 ha	*(457 /ac)*	@	1.2 £/tree		1372		*(549)*
Guards and ties	1143 ha	*(457 /ac)*	@	0.85 £/tree		972		*(389)*
Labour to plant	1143 ha	*(457 /ac)*	@	2.2 £/tree		2515		*(1006)*
Drill and establish swards	1 ha		@	465 £/ha		465		*(186)*
Irrigation drip line	2700 m/ha	*(1080 m/ac)*	@	85 p/m		2295		*(918)*
Establishment costs							**30478**	*(12191)*

Sensitivity analysis

	Change in value (+/-)	Change in av. gross margin	Value range Low	Value range High	Average gross margin range Low	Average gross margin range High
Total yield	1 t/ha	443 *(177)*	2	12	-1306 *-(522)*	3126 *(1251)*
Price	0.1 £/kg	800 *(320)*	1	1.5	1913 *(765)*	5913 *(2365)*
Gradeout*	1%	72 *(29)*	50%	85%	-87 *-(35)*	2433 *(973)*

*In fruit growing this represents the quantity reaching Class I&II standard

Performance of small organic production holdings

Farming and growing on small horticultural holdings is quite different from the larger scale units that are generally reported in this Organic Farm Management Handbook. Not only are the expectations and nature of the business different, but operating on a small scale, often with high labour input, low mechanisation and marketing to several outlet types including direct to the public, means that costs, yields and returns may bear little similarity to larger holdings. This is particularly the case for smaller growers, who make up a distinctive, innovative and important sector of organic production. Organic growing provides a way into commercial growing for new entrants and often involves less experienced staff, frequently on less-than-ideal land (many of the growers surveyed are on Grade 3 land) resulting in lower yields in some situations.

The following information is specifically focused on smaller organic producers. Yield estimates are drawn from various sources including an independent survey of growers, data collected through the Organic Market Garden Data survey 2019/20 of 10 farms[12] and data collected from over 50 growers by Rebecca Laughton in 2014 and 2015[13].

In understanding the performance of small growers, it needs to be recognised that there is extreme variability between holdings which results in a wide range in yields (up to 400% plus or minus), costs, prices and overall financial performance: no two holdings are the same. The holdings are typically in the range of 1–8 ha with between 15 and 35% in fertility building green manures each year. Labour usage varies widely with some farms being well mechanised and others relying on labour for most tasks. Some farms rely on volunteers while others are paying at least the minimum wage to all staff.

Organic yields for many vegetable crops are 50–60% of conventional, due to soil fertility, pest, disease and weed control and grade outs. This is also generally true for smallholdings. Yields tend to be similar to those for organic field scale crops, but there are exceptions. Yields can be better than similar crops grown organically on a field scale, because of higher labour input, and a longer harvest period with multiple pickings. In the table overleaf, small growers' calabrese yields are greater than the larger field scale production, but carrot yields

12. Organic Market Garden Survey Alderson, R. 2019/2020 https://organicgrowersalliance.co.uk/omg-data
13. Laughton, R. (2017) A Matter of Scale: A study of the productivity, financial viability and multifunctional benefits of small farms* (Median data collected from 50 holdings in 2014 and 2015, including access paths).

SECTION 8 HORTICULTURE GROSS MARGINS

are lower. Yield variability is likely to be due to soil type, weather and general growing conditions as well as management and marketing skills. Also using market outlets with less stringent cosmetic requirements, such as direct sales, can imply that a higher proportion of the crop can be marketed. Furthermore, intensive succession planting on many small-scale holdings results in the same area of land yielding two or even three crops in a single season.

The prices shown in the table are for box scheme/farm retail outlets, but many growers will also supply retailers and wholesalers, at lower prices. Gross margins for individual crops grown at smaller units are not provided. Variable costs data are rarely available for small multi-enterprise units, and the single crop gross margin does not appear to be a useful tool for small-scale growing. The gross margins for some field scale vegetable crops in this section and green manures in Section 9 should be referred to where this level of detail is needed for budgeting or benchmarking.

There is also no systematic data collection for the whole farm business for smaller growers. There is little information on the overall financial performance of small grower holdings, but one older survey found an average turnover of approximately £26,000/ha, from a range of £13,000 to £200,000/ha of crop grown. A more recent survey by Organic Market Garden Data found an average turnover of £90,000/ha for smaller scale growers, with staff levels of 1.8 FTE /ha and staff costs being 60% of turnover. Experience shows that businesses of 5 to 20 hectares can be viable in the long term, although management skills need to be high and marketing to the higher value outlets is important for a reasonable level of profitability. Margins and returns to the grower are often low and are sometimes dependant on the use of voluntary labour. Consequently, while they may be viable for an individual grower, the returns may not be sufficient for units where all labour and management are paid, unless there are particularly good internal markets for produce (such as supply to a farm shop) or there are other profitable elements to the business such as care farming. ORC has developed a *Horticulture Costings Tool* for small growers. This helps calculate labour, machinery and other fixed costs, and returns from the business. Further information and access to the tool is available at: https://tinyurl.com/ORC-hort-costs

Small scale horticulture yields and prices

	Yield a) - kg/m² or head/m²	Prices £/kg or £/head b) Box scheme /farm retail	SA prices charged by wholesalers c)
Potatoes maincrop	2	1 - 1.7	0.5
Brussels	1.5	2.8	
Cabbage, white (per head)	2.3	1.6 - 2.8	
Calabrese	0.9	2.5 - 3.5	
Swede	4	1 - 1.5	0.75
Cauliflower (per head)	1.65	1 - 1.4	
Leeks	2.2	2 - 3.5	2
Onions white	2.4	1.2 - 1.4	2.8
Beetroot	2.3	1 - 1.6	0.85
Carrot	2	1.3 - 1.7	2.2
Parsnip	2.2	1.3 - 1.5	0.85
Squash	2		
Broad beans	1.4		
Tomatoes	3.5		
French beans	1.5	5.5	
Leaf beet and chard	2	5.5	
Kale	1.4		
Salad leaves	1.8		
Courgette	3	2.0 - 3.0	2.9
Garlic fresh	1.1		
Spinach	0.8		
Lettuce (head)			1.21

a) Yield estimates are drawn from the following sources: Independent grower surveys for 2022; Organic Market Garden Data, Alderson, R. 2019 ; Laughton, R. (2017) A Matter of Scale: A study of the productivity, financial viability and multifunctional benefits of small farms* (Median data collected from 50 holdings in 2014 and 2015, including access paths). Yields are produce sold, kg/m² or head/m² where specified. 1kg/m² = 10 tonne/ha, including tracks.Yields vary very widely according to soils, skills and weather - yields are given for guidance only.
b) Prices £/kg (or per head where specified) are the best estimate of average prices achieved, in season, during 2022, based on information provided by commercial growers and market reports. Prices vary widely according to demand, quality, supply and volumes - prices given for guidance only.
c) The SA weekly price information provides an indication of the range of prices being charged by wholesalers to retailers, reference Soil Association Organic Market Information[14].

14. www.soilassociation.org/farmers-growers/market-information/price-data/horticultural-produce-price-data/

Forage and catch crops

Forage legumes[1]

Herbage legumes form the cornerstone of organic crop rotations because of their contribution to biological nitrogen fixation and organic matter accumulation. They are adapted to a wide variety of locations and their high nutritional value relative to grass means they can be used profitably by livestock. White clover/ryegrass mixtures form the basis of most medium- to long-term grazing swards, while high-yielding, short-term red clover/Italian ryegrass mixtures and lucerne are frequently used for forage conservation and for the fertility-building phase in arable and stockless rotations. There is widespread use of more diverse, multi-species variations on these mixtures, often including herbs such as chicory and ribgrass. Improved pasture in most of the UK is capable of growing over 12 tonnes of dry matter per hectare (t DM/ha), but the current UK average is around 8t DM/ha[2]. There is limited research on grassland productivity but more farmers are now monitoring productivity with plate meters and similar tools.

Forage yields from organically managed grassland should not be substantially different to conventional systems (see Introduction Section 1). pH, Soil type and structure, soil phosphate and sward clover content, site class and rainfall will all influence dry matter forage yields. Red clover swards are usually more productive than long-term leys and permanent pastures, achieving yields of on average 10tDM/ha (range 9-15t/ha). Grassland outputs may be about 10% less than conventional, depending on past fertiliser use.

In established organic systems stocking rates (1.4 to 1.8 LU/ha) are typically about 25% lower than in comparable conventional systems with lower stocking rates often experienced during conversion. Reduced stocking rates are a consequence of lower yields during establishment of short-term leys and reduced concentrate feeding rather than reduced grassland yields.

Using legumes, either alone or in mixtures with grasses, is increasingly recognised as offering the potential to mitigate Greenhouse Gas (GHG) emissions by reducing N_2O losses. The Leg-Link project[3] studied the characteristics of individual legume species to maximise both fertility building and the subsequent utilisation of the

1. IOTA Research Reviews, Grass clover ley species and variety selection and management; https://orgprints.org/id/eprint/13547/
2. https://ahdb.org.uk/knowledge-library/improving-pasture-for-better-returns
3. Final report available at https://tinyurl.com/LegLink-final

fertility by the following crops. Species that are of particular value in terms of forage yield and nitrogen fixation are red and white clover and lucerne; species that slow down the release of nitrogen to subsequent crops include black medic, birdsfoot trefoil, sainfoin and lucerne, hence improving the yield of subsequent crops. Other species are particularly beneficial to wildlife, such as crimson clover. The project focused on fertility building in ley/arable rotations, but many of the lessons learnt apply to leys primarily grown for forage for livestock, including the benefits of more diverse mixtures with several different legume species for better stability of forage yields and for pollinators, as well as the importance of including several grass species to maximise yield and minimise leaching[4].

The work of Rothamsted[5] shows that use of white clover results in a 40% reduction of nitrous oxide emissions compared to grass leys.

Herbs

Chicory, plantain, salad burnet, yarrow and sheep's parsley can potentially contribute to enhanced mineral uptake, the control of parasites, increased plant diversity, and drought tolerance. However, their high seed costs and poor persistence under some management regimes mean that some herbs may be better sown as strips within a field rather than as part of the ley mixture. In many situations, chicory is being grown very successfully in grass-clover mixtures for out-of-season quality grazing, its deep tap root, mineral content and anthelmintic properties.

Seeds and mixtures[6]

The retained EU regulations for organic production stipulate that organic seed must be used wherever available. Given the limited availability of organically produced seed for forage crops across Europe, the UK has specified a minimum of 70% organic for mixtures. Derogation from the certifier is still needed but can be applied for retrospectively. For seed mixtures with less organic seed, farmers will need to seek **derogation** from their certifier **before** purchase, with approval given only in extreme circumstances where it can be demonstrated that no suitable organic seed is available. Prices shown in this section are based on the price for organic seed in 2023 unless indicated otherwise. Prices for mixtures are typically based on the 70% organic.

---

4. Döring T, Howlett S (2013) Manifold green manures - Part 1: Sainfoin and birdsfoot trefoil. Organic Grower 22:34-35; Döring T, Boufartigue C, (2013) Manifold green manures - Part 2: Alsike and crimson clovers. Organic Grower 23:28-29. Available from https://tinyurl.com/Manifold-GM
5. www.rothamsted.ac.uk/news/grazing-experiment-shows-ipccs-own-estimates-climate-impacts-are
6. IOTA Research Review: Younie, D (2008) Grass clover ley species, variety selection and management. http://orgprints.org/13547/

In any forage production it is important to use seed mixtures that are 'fit for purpose'. Organic farms are dependent on quality forage in order to minimise concentrate feed cost while maintaining livestock performance. Using mixtures with suitable legumes, grasses and possibly herbs is essential to maximise both livestock performance and crop yield. Seed companies in the UK now offer a range of organic seed mixtures to meet most requirements, largely based on mixtures with red and/or white clovers, ranging from long term leys (including diverse multi species mixtures) to very short-term high quality and productive mixes. The choice of clover types and variety depends on the intended use. Red clover is used mainly in short-term and conservation leys. Larger leafed white clover varieties are more suited to rotational cattle grazing and silage, while smaller ones are for sheep and continuous grazing regimes. Mixing red and white clover in leys is becoming more common with the interest in diverse leys, but one should be aware that this loses the opportunity to have a pest and disease break in the rotation, particularly for red clover which is more disease susceptible. Herbs and secondary grass species are sometimes added to standard company mixtures. Less widely used legume species (e.g., black medic and Lucerne) should also be considered; such mixtures may be more expensive, but the cost can often be recouped quite quickly with improved performance.

The choice of organic seed is narrower, and prices will vary widely according to quality and availability. Varieties selected from the NIAB Recommended Grass and Clover Lists are generally preferable as independent information on performance and characteristics is available. Unfortunately, NIAB only tests varieties under non-organic, high nitrogen situations so the results should be interpreted with care.

Establishment and management

Conventional establishment practices are used, typically involving ploughing followed by secondary cultivations using discs, tines or power harrows, depending on soil type and conditions. The number of passes will depend on conditions and the need for perennial weed control. For herbage seeds, a firm, fine seedbed is usually required, while for larger seeds for green manures and catch crops, a rougher seed bed may be adequate. Spring tine or rotary cultivation alone may be adequate for green manure and forage brassicas following cereals. Sowing methods include drilling, pneumatic grass seeders and broadcasting, followed by harrowing and rolling. Leys can be successfully established by oversowing into existing swards although this method is not very reliable, being very weather dependant. Undersowing of white clover leys, for example under spring barley, can be a cost-effective way of establishing a ley in the final years of an arable rotation and to establish winter cover crops. While weed control is usually better there may be disadvantages in terms of ley establishment.

Good soil structure is essential for productive organic grassland, and particular attention should be paid to avoid poaching and compaction caused by heavy machinery or livestock during wet weather. Some costs for grassland machinery have been included in Section 12.

Grazing regime

The grazing regime[7,8] needs to consider the animal performance, weed control, survival of the desirable forage species, the impact on the soil and any environmental effects.

Set stocking (continuous grazing) is low cost and easy but produces less forage and poorer utilization, resulting in lower stocking rates.

Rotational grazing with paddocks or strip fencing is based on the need for a recovery period after a short grazing period for optimum forage growth utilisation. For example, the target heights for white clover/grass pastures grazed by cattle are: height on entry 10-15 cms (approx. 2,000–3,000 kg DM/ha), height on exit 5-8 cm, grazing duration 0.5–5 days and a rest period of 15 30 days, depending on season, longer in autumn and winter. Multispecies herbal leys may require different targets and plate meters may not be accurate.

Mob grazing[9] is a form of rotational grazing which provides longer rest periods between grazings of 40–80 days and much higher grass covers at entry of 30 cm+ and high residual grass height of 10–20 cm. The aim is to eat a third, trample a third and leave a third for regrowth. Grazing and rest periods vary widely between farms in the UK. Mob grazing may help to prolong the grazing season and with parasite control. Promoted on the basis of anecdotal experience there is no UK replicated research that shows that mob grazing is better than rotational grazing for soil health, for mitigating climate change or for animal productivity. A review of worldwide research on the environmental impact of grazing systems by the James Hutton Institute[10] found no consistent evidence on which to assess whether mob grazing is better than any other system. There are additional fencing and water costs and perennial weed control may be more difficult. It may be unsuitable for sheep grazing as they prefer short, high-quality forage.

---

7. Comparing grazing methods: https://tinyurl.com/comparing-grazing
8. Planning grazing strategies AHDB: https://ahdb.org.uk/knowledge-library/planning-grazing-strategies-for-better-returns
9. Mob grazing methods: https://nora.nerc.ac.uk/id/eprint/535429/
10. Climate-positive farming reviews James Hutton Institute 2022, Unravelling terminology and impacts of rotational grazing Fielding 2022 https://tinyurl.com/JHI-rot-grazing

Crop nutrition

Farmyard manures and slurry are normally applied to forage crops in the rotation, as these crops often lead to greatest nutrient off-take, particularly when cut for hay or silage. Permitted mineral fertilisers such as rock phosphate and lime may also be applied to these crops, particularly before establishment of medium/long term leys. The costs of these fertilisers have been included as a standard annual charge (see Section 6), but care should be taken to avoid double-counting where forage crops/green manures are double cropped.

Weed control[11]

Good pasture management involves maintaining the sward in good condition by cultural methods, in particular reducing weed intrusion by chain-harrowing in spring and topping strategically during the growing season. Perennial weeds such as docks and creeping thistles may be more intractable, requiring careful management of existing swards and intensive cultivations when the new ley is established, and possibly hand-pulling in established grassland. Rotavating to 7.5cm (3 inches) in July could provide effective control of docks where a 3- or 4-year ley is being incorporated before a winter cereal. Effectiveness can be further increased by repeated cultivation at approximately 2-week intervals, to exhaust the dock root reserves, and a final ploughing. This technique is highly weather dependent. For thistles, concentrating first on isolated and smaller infestations before tackling the larger ones is recommended as they have a larger potential to spread. Frequent mowing (two to three times in the first year) helps to reduce the weed population. Topping strategically can also help with weed control, and good grazing management at optimum sward heights will help to reduce the levels of weeds. Weeds in forage row crops may be controlled by mechanical inter-row hoeing as needed.

Forage conservation and silage additives

Due to later spring growth of legume-based swards, slightly later dates for turnout and cutting should be expected. Care is needed when making clamp silage from legume-based leys, in particular to ensure an adequate wilt, excluding air, sheeting overnight, double sheeting and weighting on completion. Because of the low sugars, high protein and high buffering capacity, it is essential to wilt white clover/grass and red clover/grass to 25% and lucerne to 30% dry matter as a minimum, preferably an additional 5% in each case. Although organic farmers report generally good results without the use of bacterial silage additives, there is increasing evidence that they help,

---

11. Weed control in grass and forage crops GDC Farming Connect Factsheet https://agricology.co.uk/resource/weed-control-grass-and-forage-crops/

particularly where the ideal dry matter contents are unlikely to be achieved. Use of inoculants can improve fermentation and feed value. Costs of silage making on organic farms are illustrated below.

Permanent grassland

Output/utilisation

Yield/ stocking	Mainly grazed, yields depending on soil type and fertility, sward type, and rainfall.

Variable costs

Seeds	Costs for renovation are assumed to be distributed over 10 years. The actual costs will depend on the quality of the existing sward. Lower costs if only white clover is required (see green manures for seed rates of white clover). Special considerations apply to areas that are part of agri-environment programmes.
Fertilisers	Applied on a rotational basis (see Section 6). Rock fertilisers applied according to analysis.
Other	Allow for casual labour for occasional perennial weed control.
Silage costs	See notes for leys

Permanent grassland

					£/ha	*(£/ac)*
Fertilisers					62	*(25)*
Perennial weed control					10	*(4)*
Slot-seeding/improvement	25 kg/ha	10 yearly	@	7.00 £/kg	18	*(7)*
Annual variable costs					**90**	*(36)*
Slot seeding clover only	5 kg/ha	10 yearly	@	19.00 £/kg	10	*(4)*

Leys and grassland

Output/utilisation

Yield/ stocking	Depending on type, may be grazed directly by livestock or conserved for winter feed. Herbage yields and livestock stocking rates are related to site classes, soil type and clover content of the sward (see above), for white clover on site class 2-3 approx 9-10t DM/ha; for red clover mixtures 10-11t DM/ha.

Variable costs

Seeds	Costs based on typical merchant mixtures and are shown for establishment as well as different lengths of use. Prices vary with the varieties and with the species diversity of the mixtures. Choice should depend on intended use. Add £25-37.50/ha (£10-15/acre) for perennial herb mix. A grass sward can be improved by stitching-in white clover with a specialist drill or grass harrow. Seed rate 5kg/ha at a cost of £20/kg = £100/ha
Fertilisers	Applied on a rotational basis (see Section 6). Rock fertilisers applied according to analysis.
Other	Allow for casual labour for occasional perennial weed control.

Crop management

Fertilisers	Composted FYM/slurry applications depending on availability and intensity of management. Typical FYM rates 10t/ha, applied in spring. If the grass is conserved additional manures/slurry may be applied after first cut. Avoid late application due to risk of silage contamination.
Soil aeration	Sub-soiling/spiking may be needed in medium/long term leys and permanent grassland if drainage is impeded, to allow full exploitation of soil profile by plant roots
Weed control	Chain or spring tine harrowing in spring and topping routinely for perennial weed control. Occasional hand-rogueing may be needed.

Silage making costs

Yield	14 t/ha fresh weight/cut for 2 cuts at 30%DM plus some grazing.
Silage costs	Cover average costs for sheets, wrapping or additives.
Contractor	2 cuts at 14 t/ha fresh. See Section 12 for details and charges.

Ley & grassland mixtures

		Standard ley		*Diverse mixture*		*Long-term*	
		Red or white clover				*with herb mix*	
		kg/ha	*(kg/ac)*	kg/ha	*(kg/ac)*	kg/ha	*(kg/ac)*
Seed rate		30.0	*(12)*	32.5	*(13)*	32.5	*(13)*
Assumed price		£7.00	/kg	£7.00	/kg	£8.00	/kg
Variable costs		£/ha	*(£/ac)*	£/ha	*(£/ac)*	£/ha	*(£/ac)*
Establishment seed costs		210	*(84)*	228	*(91)*	260	*(104)*
Fertilisers		62	*(25)*	62	*(25)*	62	*(25)*
Other		14	*(5)*	14	*(5)*	14	*(5)*
Annual variable costs (excl. seed)		76	*(30)*	76	*(30)*	76	*(30)*
Net annual variable costs		£/ha	*(£/ac)*	£/ha	*(£/ac)*	£/ha	*(£/ac)*
Life of ley (years)	1	286	*(114)*	303	*(121)*	336	*(134)*
	2	181	*(72)*	189	*(76)*	206	*(82)*
	3	146	*(58)*	151	*(61)*	162	*(65)*
	4	128	*(51)*	132	*(53)*	141	*(56)*
	5	118	*(47)*	121	*(48)*	128	*(51)*
	6	111	*(44)*	113	*(45)*	119	*(48)*
	7	106	*(42)*	108	*(43)*	113	*(45)*
	10	97	*(39)*	98	*(39)*	102	*(41)*

Silage conservation costs

Estimated yield - total over 2 cuts		28	t fresh/ha	@	30 %DM
Additives	@	1.8	£/t fresh		
Plastic sheets	@	1.65	£/t		
		£/ha	*(£/ac)*		
Additives and sheets		97	*(39)*		
Contractor charges		296	*(118)*		
Net variable costs 4 year ley		128	*(51)*		
Total variable costs		**521**	***(208)***	**18.6 £/t**	

SECTION 9
FORAGE CROPS AND GREEN MANURES

Green manures/cover crops[12]

Green manures have an important role in arable and horticultural systems (see Sections 5 to 7). Legume based mixtures (N fixers) form the basis of fertility building in stockless rotations, where 1-2 year clover-based mixtures contribute also to organic matter accumulation, soil structure and weed control. Undersowing with clovers (see below) can also be used in some vegetable crops (e.g. runner beans, courgettes, tomatoes). Alternatively, crops such as mustard, forage rye, fodder rape and phacelia can be used as summer or over wintering catch crops, retaining nutrients, protecting soils from erosion and contributing to weed, pest and disease control.

Normally, green manures are not utilised by livestock, but depending on the species used, may be mulched at intervals (for weed control and organic matter/nutrient cycling) and/or incorporated prior to cropping. There may be a nitrate leaching risk depending on the green manure management, timing of incorporation, the weather and subsequent cropping.

Output/utilisation

Feed — If utilised by livestock, some financial return can be obtained. The accumulated nutrients may be more readily available to the subsequent crop, but the benefits from the organic matter produced may be different.

Seed — Green manures such as red clover in stockless organic rotations may be utilised for seed production.

Variable costs

Seeds — Seed rates and costs depend on the mixture used (see Table below).

Fertilisers — Costs for rock phosphate and lime have been allocated on a rotational basis as described on p. 103, but care should be taken to avoid double-counting in the case of catch crops.

12. For choice of species guidance see: https://tinyurl.com/Agrodiversity-toolbox and for choice of species toolbox see:https://tinyurl.com/Subsidiary-crops

Forage/green manure seed prices (£/kg) (based on 2023 prices)

	Assumed value
Grasses	
Perennial ryegrass	6.70
Italian ryegrass	4.90
Timothy	9.80
Meadow fescue	11.50
Tall fescue*	10.50
Cocksfoot	6.95
Forage rye	1.20
Legumes	
White clover	19.00
Red clover	11.10
Alsike clover*	9.60
Lucerne	14.85
Sainfoin	4.95
Sweet clover	7.85
Crimson clover	9.10
Vetches	3.90
Yellow Trefoil*	15.40
Lupins, bitter blue	3.10
Field beans	1.80
Forage peas	1.50

	Assumed value
Herbs/flowering species	
Chicory*	17.35
Salad burnet*	12.90
Sheep parsley*	15.50
Yarrow*	50.00
Phacelia	9.75
Buckwheat	4.95
Fodder brassicas/roots	
Kale*	19.60
Forage rape	4.90
Mustard	4.90
Stubble turnips	10.90
Maincrop turnips*	11.40
Swedes	62.00
Fodder beet*	70.00
Raddish*	8.1
Westerwolds	4.15

* = price based on non organic

Green manures (based on 2023 prices)

Crop	*Seed rate* kg/ha *(kg/ac)*	Seed price £/kg	*Seed cost* £/ha *(£/ac)*	*Variable costs* £/ha *(£/ac)* Including fertilser
Perennial ryegrass/	15 *(6)*			
red clover mixture	12 *(5)*		234 *(93)*	296 *(118)*
White clover	5 *(2)*	19.00	95 *(38)*	157 *(63)*
Red clover	15 *(6.0)*	11.10	167 *(67)*	229 *(91)*
Trefoil	7 *(3)*	15.40	108 *(43)*	170 *(68)*
Mustard	20 *(8)*	4.90	98 *(39)*	160 *(64)*
Sanfoin	75 *(30)*	4.95	371 *(149)*	433 *(173)*
Vetches**	75 *(30)*	3.90	293 *(117)*	355 *(142)*
Vetch/Rye	185 *(74)*	0.96	180 *(72)*	242 *(97)*
Phacelia	10 *(4)*	9.75	98 *(39)*	160 *(64)*
Forage rye	190 *(76)*	1.20	186 *(74)*	248 *(99)*
Buckwheat	63 *(25)*	4.95	312 *(125)*	374 *(150)*
Field beans	200 *(80)*	0.65	360 *(144)*	422 *(169)*
Lupins	50 *(20)*	3.10	155 *(62)*	217 *(87)*
Forage peas	100 *(40)*	1.50	150 *(60)*	212 *(85)*

** Seed range 60-125kg/ha

Refuges for beneficial insects (flowering field margins, bug banks)

Bug banks (0.3m high x 1.5m wide)

Mixture	g/m²	*Seed rate* kg/ha *(kg/ac)*	*Seed cost* £/ha *(£/ac)*
Cocksfoot	2.00	20.0 *(8.0)*	139 *(56)*
Phacelia	0.25	2.5 *(1.0)*	24 *(10)*
Buckwheat	0.25	2.5 *(1.0)*	12 *(5)*

Arable forage, forage brassicas and root crops

Output/utilisation

Feed Depends on type and intended utilisation period. May be grazed in situ (e.g. kale in autumn/early winter, forage rye in early spring) or harvested and conserved/stored for winter feeding (maize, whole-crop silage, fodder beet). Dry matter yields for whole crop silage are likely to be in the range of 8-10t DM/ha, depending on crop and soil conditions. A second cut can be achieved if combined with undersown leys.
Irrigation may be necessary for fodder beet.

Variable costs

Seeds Prices as indicated on next page. If no organic seed available, permission for the use of undressed seed from conventional sources from the Control Body is required. Fodder beet: graded monogerm assumed.

Fertilisers Potash and phosphate fertilisers applied on rotational basis.
For fodder beet there is an additional need to ensure that sodium levels are satisfactory, and an application of agricultural salt may be required, either as solid or liquid.

Other Allowance for additional manual, mechanical and/or thermal weed control costs in maize, kale, swedes and fodder beet. Contractor charges for maize and fodder beet harvesting see Section 12. Cost of cleaning fodder beet prior to utilisation should also be considered, but not included here.

Crop management

Rotation Maize: first crop after grass/legume ley
Peas: grown on lighter soils, possibly with oats/barley or in whole-crop silage.
Whole-crop silage: summer catch crop, harvested as whole-crop silage for winter livestock feed.
Fodder brassicas and roots: commonly double-cropped, e.g. before a spring cereal or following forage rye.
Fodder beet: nutrient demanding and weed susceptible crop that will need to be grown as the first crop following a ley or annual green manure.

FYM/Slurry Requirement depends on previous cropping and soil fertility.
Maize: responsive to compost or manure, e.g. 50m^3 slurry/ha.
Kale and fodder beet: 20-30t FYM/ha or application of compost at up to 25t/ha.

Weed control	General: weed strike, and during initial stages of plant growth using inter-row hoes, 1-3 passes, depending on weed competition and crop type. Maize: specialist inter-row hoes, 2-3 passes, depending on weed competitiveness. Flame weeding at 5 cm and 25 cm stages an option as maize relatively heat tolerant. Spider-tine hoes can be used to good effect by turning soil into rows and burying weeds. Undersowing (e.g. with legumes) can also help suppress weeds and provides a firm vegetation base to protect soils from harvest machinery and postharvest. Peas: weed control during initial stages of plant growth may be carried out by in-crop cultivations using spring tine harrows or inter row hoe; a dense crop will reduce weed competition. Fodder beat: High standards of weed management are essential in this non-competitive crop; including use of a weed strike, later drilling into warm soils and 4 or 5 passes of an inter-row hoe. A single hand hoeing in the row may be required.
Harvest	Kale and other brassicas strip grazed. Fodder beet is a late harvested crop; it is essential that it is only grown on free-draining soils and harvested under good conditions in order to avoid damage to soil structure. Can be lifted using root harvester and clamped for winter feed, or grazed in situ. Small areas may be hand lifted at 4 hours/t.

Arable forage crops

				Arable silage mixtures	
	Maize	*Peas*	*Rye*	*Cereal/Peas*	*Oats/Vetches*
Seed rate kg/ha (kg/acre)	50 *(20)*	100 *(40)*	185 *(74)*	60 130	125 60
Seed cost (£/kg)	3.92	0.9	0.98	0.65 0.9	0.63 3.90
	£/ha *(£/ac)*	£/ha *(£/ac)*	£/ha *(£/ac)*	£/ha *(£/ac)*	£/ha *(£/ac)*
Seeds	196 *(78)*	90 *(36)*	181 *(73)*	156 *(62)*	312 *(125)*
Fertilisers	62 *(25)*	62 *(25)*	62 *(25)*	62 *(25)*	62 *(25)*
Other	67.5 *(27)*				
Total variable costs	**326** *(130)*	**152** *(61)*	**243** *(97)*	**218** *(87)*	**374** *(150)*

Forage brassicas and roots

	*Seed rate**	*Seed cost*	*Fertilisers*	*Other costs*	*Variable costs*
	kg/ha *(kg/ac)*	£/ha *(£/ac)*	£/ha *(£/ac)*	£/ha *(£/ac)*	£/ha *(£/ac)*
Kale - broadcast	4 *(1.6)*	78 *(31)*	62 *(25)*		140 *(56)*
Forage rape - drilled	8 *(3.2)*	39 *(16)*	62 *(25)*	33.75 *(14)*	135 *(54)*
Forage rape - broadcast	10 *(4.0)*	49 *(20)*	62 *(25)*		111 *(44)*
Mustard - broadcast	24 *(9.6)*	118 *(47)*	62 *(25)*		180 *(72)*
St. turnips - broadcast	7 *(2.8)*	76 *(31)*	62 *(25)*		138 *(55)*
Radish	10 *(4.0)*	81 *(32)*	62 *(25)*		143 *(57)*
Swedes	3.5 *(1.4)*	217 *(87)*	62 *(25)*	20.25 *(8)*	299 *(120)*
Fodder beet	3.8 *(1.5)*	266 *(106)*	62 *(25)*	40 *(16)*	368 *(147)*

*variety will influence seed rate

Undersown forage

Undersowing forage crops is an important aspect of mixed cereal and pasture rotations, as most of the cultivations for establishing the crop are carried out under the preceding spring cereal crop and can achieve a timelier establishment of the clover in the autumn. A number of nurse crops can be used. In the example below, a white clover/ryegrass mixture is undersown into spring barley. Red clover is only suitable if long strawed cereals are used or if harvested as whole crop silage.

Output/utilisation

Cereal Reduced grain yield due to herbage competition (financial impact is included as variable cost in the Table). In some cases, cereal may be harvested early for whole crop silage (see below).

Herbage May be grazed by in autumn, but full production only in following year.

Variable costs

Seeds Reduced seed rate for cereals.

Fertilisers Costs included in cereal gross margin.

Undersown forage

					£/ha	*(£/ac)*
Seed - standard mix	30.0 kg/ha	*(12.0 kg/ac)*	@	7.00 £/kg	210	*(84)*
Cereal yield reduction	0.6 t/ha	*(0.2 t/ac)*	@	260 £/t	156	*(62)*
Cereal seed saved	40 kg/ha	*(16.0 kg/ac)*	@	600 £/t	-24	*-(10)*
Total variable costs					**342**	*(137)*

Further reading

Döring T et al (2013), Using legume-based mixtures to enhance the nitrogen use efficiency and economic viability of cropping systems. http://orgprints.org/24662.

Younie D (2012) Grassland Management for Organic Farmers. The Crowood Press, Marlborough.

Website and technical guides

https://agricology.co.uk/resources/?_resources_farming_themes=crops-forage-horticulture

https://agricology.co.uk/resources/?_resources_farming_themes=weeds-pests-and-diseases

https://ahdb.org.uk/knowledge-library/ahdb-grass

The Cover Crops and Living Mulches Toolbox (ORC/EU Oscar project) www.agricology.co.uk/resources/agrodiversity-toolbox

ORC Research Reviews:

www.organicresearchcentre.com/resources/publications

- Grass clover ley species and variety selection and management: https://tinyurl.com/IOTA-grass-clover
- The role of management of herbal pastures for animal health, productivity and production quality: https://tinyurl.com/IOTA-herbal
- The role and management of whole-crop forage for organic ruminants: https://orgprints.org/id/eprint/13556/

Clover Management Guide IBERS: www.rwn.org.uk/aber-clover-management-guide.pdf

Rosenfeld A and Rayns F (2011) Sort out your soil– A practical guide to Green Manures, 2nd ed. Garden Organic and Cotswold Seeds. https://tinyurl.com/Sortoutyoursoil

Farming for a better climate: Comparing Grazing Methods www.farmingforabetterclimate.org/resource/comparing-grazing-methods/

Clover Management Guide IBERS www.rwn.org.uk/aber-clover-management-guide.pdf

Role of livestock on organic farms

Livestock are integral to most organic farming systems, particularly those with permanent pasture and ley/arable rotations. The only exceptions are stockless horticultural and arable farms resulting from specific moral or practical decisions made by the farmer, such as vegan organic systems, or where there is an absence of farm infrastructure to support livestock farming. Ruminant livestock are able to utilise and provide a financial return to the leys with legumes and other herbage species that contribute to nitrogen fixation and the fertility-building phase of the rotation. They can utilise cellulose and hence energy from herbage that would not otherwise be available for human consumption and the return of livestock manures provides a targeted nutrient source for subsequent crops. Livestock thus play an important role in nutrient and energy cycling. They can also contribute to weed, pest and disease control through grazing and forage conservation. As with crop production, an emphasis on single species can lead to problems, including internal parasites or weeds (e.g., bracken in permanent pasture). The mixing of species, such as sheep and cattle or sheep and poultry, can contribute to the control of parasites and improve grassland management.

Principles of organic livestock production are clearly stated in the EU Regulations and include the preferences for organic inputs (feed and young stock) and restriction of external inputs, choice of adapted breeds and encouraging natural immunity of the animal. The organic standard requirements are largely harmonised across the EU but national requirements (e.g., Farm Animal Welfare codes) apply. Following the UK exit from the EU, the EU Regulation 834/2007 has been retained in Great Britain, as laid out in The Organic Production and Control (Amendment) (EU Exit) Regulations 2019, No. 693. Since then, EU Regulation 2018/848 has broadened the scope of EU Regulation 834/2007 and reviewed the rules for organic livestock production and production requirements for new species. Producers considering the export of animal products to the EU, EEA, and Northern Ireland should ensure that they meet the export requirements for doing so, detailed here:

- www.soilassociation.org/media/21977/exporting-guidance-for-gb-to-ni-or-eu.pdf

Organic livestock enterprises should be land-related and site-based, and the number of livestock must be appropriate for the area available in order to avoid problems with internal parasites, overgrazing and environmental pollution. The housing and husbandry should observe a high level of animal welfare

respecting the specific behaviour of each species, considering sex and age. With some exceptions organic stock should be fed a diet based on home-grown and organic feed. This effectively excludes intensive pig, poultry and cattle production, which depend heavily on bought-in feeds and lead to livestock 'waste' disposal problems. The manures produced should be capable of being absorbed by the agricultural ecosystem without leading to disposal or pollution problems. Stocking rates should reflect the inherent carrying capacity of the farm and not be inflated by reliance on 'purchased' hectares in the form of imported feeds or exported 'waste' material.

All organic animals must have access to open air areas and preferably pasture whenever the weather conditions and state of the ground permit. If access to open air for organic poultry has to be restricted to protect public or animal health, organic status can be maintained if the birds have access to roughage and other suitable material in order to meet their ethological needs.

Origin of stock, breeding, rearing and weaning

EU Organic Regulations require that all livestock sold as organic are born and raised throughout their lives on an organic holding, with the exception of animals introduced for breeding (10% for ruminants, 20% for other mammals), animals present on the farm at the beginning of conversion and poultry, for which exceptional rules apply. In addition, all certification bodies require that such converted animals should not be used in organic meat production.

Dairy cows that are on a farm prior to conversion can be converted over a 6-month conversion period. Breeding males and replacement females, subject to disease status checks, may be brought in from non-organic sources as part of the 10% non-organic livestock replacement allowance for cattle and 20% for sheep, goats and pigs. Emphasis is placed on maintaining closed herds and flocks, i.e. breeding replacements on the farm, so as to minimise the risk of importing diseases from elsewhere and in order to develop stock that is well suited to the conditions on a particular farm.

Specific breeding objectives for organic systems include high lifetime yields from forage, fertility and resistance to mastitis in dairy cows, suitability for outdoor systems (e.g., hardiness) for pigs, and internal parasite resistance in sheep. Breed choice or cross breeding will also be related to the quality requirements of target markets and any other specific requirements of the system. The use of AI is permitted by organic standards, but not embryo transfer[1].

1. Dairy cow breeding for organic farming https://orgprints.org/5975/

Calves have to be fed on organic whole milk or organic milk replacer (as a limited percentage) up to the age of 12 weeks. Multiple suckling by a nurse cow is often preferred in dairy herds but rearing with 'calf at foot' may also be an option for some farms (a practice which is on the rise particularly with smaller herds who are direct selling dairy products). In some cases, e.g., suckler beef production, natural weaning may also be practised. Although these practices add to production costs, it is often argued that the benefits in terms of subsequent health, production and longevity more than compensate.

All stock intended for meat production has to be born on an organic farm. The only exception is poultry production, where conventional chicks less than 3 days old can be reared with prior approval from an inspection body. Organically reared livestock can be sourced through the Soil Association's Organic Market Place[2] and Organic Farmers and Growers[3].

Currently non-organic replacement pullets can be used for laying units, provided they are reared to the organic feed and veterinary standards and are brought onto the organic unit at less than 18 weeks old. The EU Commission regularly review the ability to request a derogation for the use of non-organic pullets and the current deadline has been extended to 2025.

Livestock nutrition[4]

The aim in organic livestock feeding is to rely primarily on home-produced feeds suited to the animal's evolutionary adaptation. Avoiding the use of feedstuffs that are suitable for direct human consumption is part of the wider debate about the role livestock should play in the global food system. At least 60% of the feed for ruminants and at least 20% (in GB and now 30% in EU) of the feed for pigs and poultry must come from the holding itself, or if this is not feasible produced in cooperation with another organic farm in the same region.

Ruminant rations should be predominantly forage-based (at least 60% of the diet). This is normally comprised of grass/clover or herb-rich grazing and conserved forage, although there is clearly potential for other forage crops such as arable silage, fodder-beet and maize. The use of cereals and pulses in ruminant feeding (up to a maximum of 40% in the diet) is aimed at balancing

2. www.soilassociation.org/organicmarketplace
3. www.organicfarmers.org.uk/classifieds
4. Organic beef and sheep nutrition. IOTA Technical Leaflet. Available here: https://tinyurl.com/IOTA-beef-sheep

the diet rather than stimulating additional production. However, this may be increased to 50% for the first three months of a dairy cow lactation. Pig and poultry rations will inevitably rely on cereals and pulses, although there is also a need to provide some green material and indeed considerable reliance can be placed on forage crops in pig diets[5].

European organic regulations aim to achieve fully organic diets for all animals. Ruminants are required to have a 100% organic diet, but in the case of pigs and poultry there is an EU Commission organic standards allowance, upon application, to feed up to 5% non-organic proteins, recognising the ongoing difficulties in sourcing sufficient, good quality organic protein. This allowance has been extended to the end of 2025.

The non-organic limit does not apply to fishmeal and other products and by-products of sustainable fisheries, which is classified as a non-agricultural ingredient and therefore can be included in organic pig and poultry diets providing it is listed in Annex V (Article 22/EC 889/2008). However, feed mills that also produce cattle diets are not allowed to handle products of animal origin.

Any operator using non-organic feed must demonstrate that organic feed was not available and that the use is in line with the principles of organic farming (Article 4 and 5/EC/ 834/2007). A specified list of permitted conventional feeds excludes GM ingredients and the use of synthetic amino acid and certain types of vitamins. Synthetic vitamins A, D and E identical to natural ones can be used for ruminants.

Trace elements and mineral supplements are not fed routinely. However, basic minerals that are listed as feed materials can be fed to stock (e.g., in the form of natural or chelated minerals, rock salt or seaweed meal and natural or synthetic (monogastrics only) vitamins) as long as these are on the approved list and there is a demonstrable need. Please contact certification bodies for details. Problems such as hypomagnesaemia are less common on organic farms because of reduced fertiliser use, but pastures may be dusted with calcined magnesite or magnesium supplements fed if necessary.

Up to 30% of the organic proportion may come from land in conversion. This proportion can increase to 100% if the feed is produced on the holding itself but only 20% in the first year of conversion.

5. ICOPP Technical Notes:
Crawley et al (2015) Fulfilling 100% organic poultry diets: Roughage and foraging from the range (No 2) ORC. http://orgprints.org/28090/;
Crawley et al (2015) Fulfilling 100% organic pig diets: Feeding roughage and foraging from the range (No 4) ORC. http://orgprints.org/28088/

Dairy cow rations[6]

Dairy cows should be fed mainly on home-grown forages from grass/clover leys, herbal leys, permanent pasture and other suitable forage crops (at least 60% of the ration dry matter must be provided by forages), supplemented by organic straights or concentrates. 100% of the diet must be from organic sources. The diet should be structured to maintain a well-balanced rumen to promote efficient forage utilisation. Diets may be supplemented with minerals and micro nutrients to correct imbalances in home grown forages.

The main challenge in feeding dairy cows is to encourage high dry matter intake immediately prior to calving and during the first three weeks of lactation. Ensuring good intakes of high-quality forage will help to provide enough energy in early lactation. In this period, it is especially important to offer high-quality forage ad libitum. The combination of high-quality conserved forages and well managed grazing will reduce reliance on purchased feeds. Well-managed grass/clover leys providing *ad libitum* grazing should not require much supplementary feeding, except where forage quality is deteriorating or if there is a need for buffer feeding with whole-crop to provide high energy fibre. During the winter-feeding period, the level and efficiency of milk production is driven by silage quality; ensuring early cut high energy first cut and subsequent cuts of high protein silage is crucial.

Good forage management is essential and the benefits include reduced concentrate usage, improved health status and the potential to increase longevity. It is possible to achieve over 4,000 litres of milk from forage.

Protein levels of grass/clover mixtures are variable throughout the year. Grass/clover leys can produce high quality silage, (e.g. exceeding 11 ME, 18% CP) and can be very high in crude protein (in excess of 25%). There may be times when supplementary feeding (for energy and/or protein) is needed to ensure that the animals' requirements are met. Forage analysis should be undertaken in the early and late season in particular. Buffer feeding can provide additional fibre and can reduce the risk from bloat.

Appropriate ingredients, such as cereals and pulses, need to be selected for supplementation to balance the low pH, high rumen degradable protein, and low fermentable energy of silage. The use of whole-crop, fodder-beet, kale and maize silage can also be worthwhile. Mineral analysis of forages and

6. Organic dairy cow nutrition IOTA technical leaflet 1: https://tinyurl.com/IOTA-dairy-nut

home-grown feeds should be undertaken to ensure that the correct level of supplementation is carried out. Several companies are producing compound feeds that comply with organic standards.

Cows show many signs that help understand their health, well-being, nutrition and production. The challenge for the dairy farmer is how to interpret these signals and use them. Several tools have been developed to assist this. The Obsalim® cards method[7] aids the diagnosis of the metabolic imbalances in cattle which can also be combined with a guidebook and computer programme. Cow Signals[8] offers books and training courses at various levels though some solutions offered are tailored towards conventional management.

The tables on the next pages provide some guidance on nutritional content and price range of a number of suitable feedstuffs for organic animals and sample rations for dairy cows. The price and availability of organic feeds has varied widely over the last few years and the availability of some imported and home-produced feeds has become limited or unavailable. However, it has to be stressed that the nutritional content of forages varies considerably, so specific ration planning based on regular forage analysis should be carried out. All feeds other than straights must be sourced from approved organic feed suppliers.

7. www.obsalim.com/en/index.htm
8. www.cowsignals.com/books/

Livestock feed nutritional value and prices

Nutritional values per kg dry matter	*Dry matter*	*Metab. energy MJ*	*Dig. Cr. protein g*	*Crude protein*	*Fibre %*	*Price range £/t (fresh weight) delivered bulk*		
						min	*max*	*Assumed*
Organic forage/fodder component								
Wh. Clover/grass grazed	16%	11.0	150	14-28%	*14%*	See Section 9 for costs		
Wh. Clover/grass silage	25%	10.5	115	14-16%	30%	See Section 9 for costs		
Wh. Clover/grass hay	85%	9.0	58	10%	30%	See Section 9 for costs		
Rd. Clover/grass silage	30%	9.0	135	14-18%	30%	See Section 9 for costs		
Lucerne silage	33%	8.5	115	17%	30%	See Section 9 for costs		
Maize silage	30%	10.8	70	9%	23%	See Section 9 for costs		
Fodder beet	18%	12.5	55	7%	6%	See Section 9 for costs		
Kale	14%	11.0	115	15%	18%	See Section 9 for costs		
Cereal/ pea whole crop silag	38%	11.2	70	12%	26%	See Section 9 for costs		
Potatoes (chats)	21%	12.5	47	10%	4%	10	35	25
Organic concentrates component								
Cereals	86%	13.5	85	11%	11-23%	315	420	320
Cereals (in conversion)	86%	13.5	85	11%	11-23%	295	400	300
Field beans	86%	12.8	230	27%	21%	400	600	440
Peas	86%	12.8		24%	19%	445	625	470
Lucerne meal	90%	10.0	130	18%	47%	340	380	380
Lupins	88%	13.8	280	36%	12%	n/a	n/a	320
Rape Expellers (30% proteir	94%	13.5		32%	10%	625	1050	645
Soya Expellers	92%	15.0	380	46%	7%	675	1050	720
Full fat soya	90%	18.7		40%	6%	n/a	n/a	550
Sunflower Expellers	94%	12.9		32%	20%	550	850	570
Maize	88%	14.1		9%	4%	320	450	350
Compound feeds (100% organic)		Approved suppliers only						
14% Protein	88%	12.8	14	16	10	520	595	535
16% Protein	88%	12.8	16	18.2	10	530	615	555
18% Protein	88%	12.8	18	20.5	10	545	620	580
22% Protein	88%	12.8	22	25	10	585	690	635
Minerals and trace elements								
Seaweed meal	86%	8.8	73	14%	10%	1200	1550	1375
General purpose (8% Phosp	97.5					300	450	375
High phosphorus (12%)	97.5					415	550	475
High magnesium						390	540	455
Liquid magnesium (5% Mg syrup) per 500 litres								280
Trace elements (e.g. Cu, Se, Co) per 25 litres								120

All prices quoted as 2023 assuming a min. of 1t drop and incl delievery.

Sample rations for dairy cows (per cow per day in kg fresh weight)

Dairy Cow ration examples

Example No	1	2	3	4	5	6	7	8	9	10
Target milk yield (litres/day)	17	25	30	18	25	30	25	25	30	35
Live weight change	0.1	0.1	0	0.5	0	0	0	0	0.2	-0.2
Live weight	525	525	625	600	625	625	625	625	625	625
ME (FiM) requirement	159	201	233	188	208	233	203	203	242	248
Organic forage/fodder Component (kg per day freshweight as fed)										
Wh. Clover/ grazed grass	65	75	65							
Wh. Clover/grass silage				45	55	55			50	
Wh. Clover/grass hay	1	1	1			1			1	
Rd. Clover/grass silage							32	40		30
Lucerne silage										
Cereal/ pea whole crop silage								10		12
Maize Silage							16			
Fodder beet									15	
Kale				15						
Potatoes (chats)										12
Organic concentrates component										
Cereals			5		2	2	2	3.25		2
Cereals (in conversion)										
Field beans							2	3.5		3
Peas					2	1.5	2	1.5		1.75
Lucerne meal										
Lupins										
Rape Expellers (High protein)										
Soya Expellers								0.5	1	
Full far soya										
Sunflower										
Compound feed (100% organic)										
14% Protein		4								
16% Protein			5	3						5
18% Protein										
22% Protein						3			7.5	
Conc. use/cow (kg DM)	0	3.44	8.6	2.58	3.44	5.59	5.16	7.53	7.31	10.1
Dry Matter Intake (Kg)	10.7	15.7	19.2	16.1	17.2	19.6	18.5	20.8	22.3	23.4
M/D required (Mj ME / KG DM)	14.93	12.81	12.14	11.69	12.13	11.92	10.95	9.74	10.85	10.59
Crude protein (%)	18.0	18.0	17.0	18.0	16.5	17.0	17.3	17.3	16.5	17.5
Concentrate Cost (£/cow)	0	2.14	4.38	1.67	1.58	3.25	2.46	3.65	5.48	5.56
Conc. cost per litre (p/ l)	0	8.56	14.6	9.25	6.32	10.8	9.84	14.6	18.3	15.9
Conc. price (£/t)	0	535	438	555	395	500	410	417	645	473

The Dairy Cow Ration table provides some examples of suitable rations to optimise both production from forage and cow health. The first grazing ration (Ration 1) is a forage ration for once-a-day milking at peak production.

Rations 2 and 3 are for spring calving different cow species. Ration 4 is based on mid-lactation at 200 days in milk. Rations 5 and 6 are for a housed early lactation with Ration 5 using a total mixed ration (TMR). Ration 7 includes the use of alternative silages. Ration 8 is a mid-lactation ration with the use of a whole crop silage. Rations 9 and 10 are robot type diets with all cows in the parlour for Ration 9, and the use of TMR in Ration 10. Most rations include the use of cereals, preferably rolled or crimped. The main emphasis in all these diets is to provide sufficient energy, to avoid feeding excessive protein (difficult during the later grazing season) and most importantly to provide physically effective fibre.

Animal welfare and housing[9]

Organic production standards contain animal welfare provisions that are an important component of successful organic management, influencing health, behaviour and performance. An understanding of basic animal welfare considerations is therefore fundamental to the interpretation and implementation of production standard requirements. Organic production standards set out the baseline conditions in which animal welfare conditions are to be met with wider aims to work towards achieving welfare goals according to the organic principles[10].

Mutilations should not be carried out routinely, but the control body can authorise individual exceptions where not doing so risks a worsening of animal welfare or operator safety. Defra has published guidance on how these exceptions are to be authorised, which are also relevant to organic farming[11]. Suffering must be avoided by appropriate anaesthetics for dehorning and other similar operations, as specified in the welfare codes. The use of pain killers should also be considered for associated post-operative pain.

All organic animals must have outdoor access and housing appropriate to animal welfare and behavioural needs under organic production standards. Battery cages, tethering, fully slatted floors and systematic mutilations are prohibited, with the emphasis instead on free-range systems, particularly for pig and poultry production. The continual housing of pigs and sheep is not permitted during any stage of production. The exceptions being during times of notifiable disease (e.g., avian influenza), and in wintertime to preserve animal welfare and pasture quality. For beef cattle continual housing is only permitted if there are safety issues, e.g., with bulls. The exercise area for pigs must allow for the separation of dunging and rooting activity.

9. Poultry Management, IOTA Research Review: https://tinyurl.com/IOTA-poultry-nut
10. www.ifoam.bio/why-organic/shaping-agriculture/four-principles-organic
11. https://www.legislation.gov.uk/uksi/2007/1100/contents/made

Where livestock must be housed, appropriate social grouping is required, and the standards specify minimum space allowances for different livestock species and ages[12]. There is some variation between UK standards, i.e., the retained EU regulation, and between UK certification bodies. This is because certification bodies may raise their standards higher than what is required according to regulations.

There are several programs for animal welfare assessments available, one of the biggest and most recommended being Welfare Quality[13]. These assessments require time and diligence to complete but provide a helpful overview of herd health and progress. New technological developments can also allow for easier on-farm monitoring of animal health, e.g., using an intra-ruminal temperature monitoring bolus for early detection of body temperature changes in cattle[14]. Overall, reduction of disease through monitoring approaches is important, but the overall goal should focus on positive welfare as health is more than the absence of disease. Reducing the number of days with disease is desirable but increasing the overall resilience of the herd should be the aim.

Livestock health

Animal health promotion in organic farming is based on preventive management and good husbandry. The aim is to maintain a high health status of the animal and to minimise disease pressure and stress. For this a good stockperson and attention to detail are needed, as well as optimising breeding, rearing, feeding, housing and general management (including bio-security) in order to achieve stability and balance in the farming system. All organic livestock farms are required to have a written animal health plan for certification. The plan should be drawn up with the help of a specialist organic advisor and a veterinary surgeon (costs approximately £300-500 initially, depending on circumstances), and must be updated annually. There is now funding available under the Animal Health and Welfare Pathway[15] for an annual funded vet visit, which covers the vet's time and some diagnostic tests. Payment rates are:

- £684 for pigs
- £436 for sheep
- £522 for beef cattle
- £372 for dairy cattle[16]

12. https://www.soilassociation.org/media/23378/gb-farming-growing.pdf
13. http://www.welfarequalitynetwork.net/en-us/reports/assessment-protocols/
14. https://www.agrismart.co.uk/farming-solutions/bolus/
15. https://www.gov.uk/government/publications/animal-health-and-welfare-pathway
16. https://tinyurl.com/Defrablog-AHW-review

Animal health and welfare plan templates should be available from your certification body, marketing company, quality assurance scheme or in Wales through local government offices. All health plans, however, should contain the following key areas: identifying the current health and welfare status on the farm, prioritising areas and key performance indicators to be addressed in the short, medium and long term and setting targets; devising a strategy for achieving goals; collecting data/information to evaluate success or failure of strategy and finally a regular review of progress (usually annual). This process is repeated over time to continually improve health and welfare on the farm.

Preventive medication is restricted to risk assessment-based use of vaccination, herbal and homoeopathic remedies and nosodes for known farm problems or strategic use of other medicines (excluding antibiotics) in the context of a health plan. Growth promoters and the routine use of hormones and antibiotics (e.g., prophylactic dry cow therapy) are not allowed. Where possible, treatment of ailments is approached by supporting the animal's disease resistance and recovery through complimentary therapies, such as the use of herbal remedies, reduction of stress, reducing pain, physical treatment and appropriate feeding. Homoeopathy is permitted[17]. The website www.farmhealthonline.com provides useful information on disease management and welfare. Homoeopathy at Wellie Level (HAWL)[18] offers a course to gain homoeopathic skills, specifically designed for farmers and stockpersons.

Conventional disease treatment should be used in all cases where it is necessary to prevent prolonged illness or suffering. However, an individual animal that receives more than three courses of conventional treatment within a year may lose its organic status. Conventional treatments also incur longer withdrawal periods than stated by the manufacturer. Records of all cases of ill health and treatment (including use of complementary medicines) must be made available at inspection. Some strategies for prevention and control of animal health problems including costs of some frequently used treatments are outlined also in this section.

UK organic standards on antibiotic use ensure animal welfare while minimising food residue risks. However OMSCO (now Organic Herd) has developed export markets for British milk and cheese by prohibiting the use of antibiotics,

---

17. Hansford, P. (1992) The Herdsman's Introduction to Homoeopathy. Helios Homoeopathic Pharmacy, Tunbridge Wells;
    Day, C (1993) T A Guide to the Homeopathic Treatment of Beef & Dairy Cattle Daniels.
18. 3-day course in Homeopathy at Wellie level. www.hawl.co.uk/

amongst all its suppliers, in order to meet the US NOP organic standards which are more stringent in this one respect. It should be carefully considered if it is worthwhile to become antibiotic-free, if the general herd health is suitable and which alternatives to antibiotic treatment are available.

Mastitis

Dairy parlour and housing maintenance and hygiene, along with good milking practice, provide the main means of control. Early detection can allow successful treatment by frequent stripping of the affected quarter and cold-water massage and/or the use of licensed herbal udder creams. Homoeopathic remedies and nosodes may be used; while there is very limited scientific evidence of their efficacy there is considerable anecdotal evidence that homoeopathy is effective as part of a health management plan. In severe cases, recourse can be made to antibiotics subject to extended withdrawal periods. As antibiotic dry cow therapy is not permitted on a whole herd/prophylactic basis, special attention should be paid to minimise stress during the drying off period along with good hygiene in dry cow environments. To avoid inefficient use of antibiotics, the type of bacteria that caused the mastitis can be identified with rapid on-farm test kits such as VetoRapid[19]. For gram-negative pathogens, certain antibiotics are proving ineffective, whereas gram-positive infections can still be treated successfully.

Orbeseal, an internal teat sealant is effective at preventing new infections during the dry period (often a problem in organic herds without systematic antibiotic use). It is not a treatment for mastitis. Orbeseal is categorised as a prescription only medicine (POM) product and care should be taken to insert it using the strictest hygiene procedures.

There is increasing interest in selective antibiotic treatment of clinical mastitis following on farm culture or a similar test, as some pathogens have a high spontaneous cure rate. This should be carried out under veterinary advice.

19. https://vetgrad.com/clickthrough.php?sourceID=1925&type=link

Costs of mastitis treatments

	Details	Price	Cost/treatment
Uddermint	6 x 10ml/treatment	£21/600 ml	£1.26/treatment
Orbeseal	Internal teat sealant for drying off		£6-8/cow
Antibiotic tubes		3-5 tubes per case	£ 3-4 per tube
Homoeopathic nosodes Potency C30 (preventive applications)	0.5ml/cow/month for 3 months	£73/100ml	£73/100ml
Homoeopathic remedies According to symptoms (treatment)	3-9 x 1ml doses (diluted)/treatment	£78/100ml	Average £4.00/ course

N.B. Costs above are illustrative and may vary considerably depending on source of purchase and advice given for administration. With all these treatments, the period for withholding milk from sale can be substantially reduced compared with antibiotic use, leading to potential savings.

Foot problems[20]

Control should be based on gentle handling, appropriate feeding, bedding and flooring (including trackways), with good housing design (avoiding steps and narrow alleys/corners) and hygiene and supplemented with corrective foot trimming when required. Special care should be taken when introducing newly calved heifers to the milking herd. Zinc and copper sulphate (£60/20kg) footbaths are permitted (although the legality of copper sulphate footbaths is debated as there is concern about the build-up of residues) and iodine sprays may be used. Homoeopathic nosodes can also be used as a preventive measure, although scientific evidence for their efficacy is lacking. Antibiotic sprays are effective against bovine digital dermatitis, but will incur a 48-hour milk withdrawal. Alternative products to antibiotic sprays are also available, and some have been evaluated in controlled trials, e.g., Intra Hoof Fit Gel.

External parasites

External parasites, by their nature, are difficult to control through management strategies and treatment is often needed. 'Nuisance' flies represent a hazard and sometimes are a serious problem for most categories of livestock. Various approaches can be used to reduce the fly problem in the case of cattle, including field and farm-yard hygiene and the introduction of predatory wasps to control

20. https://ahdb.org.uk/knowledge-library/introduction-to-the-healthy-feet-programme

breeding sites, good ventilation, water spray exclusion barriers and traps in the dairy parlour, repellents such as citronella oil, garlic or Stockholm tar and barriers such as Collodion to protect dry cows. Some pesticides are permitted, but are subject to varying withdrawal periods under organic standards which may make their use impractical for milking cows.

Permitted control products for external parasites

Problem	Active ingredient	Price	Cost/treatment
Sheep scab	Doramectin	£70/200 ml	£0.80/ewe
Fly strike (sheep, prevention)	Dicyclanil	£230/5 litres	£2.00/head
Fly control (dry cows)	Deltamethrin	£81/1 litres	£0.81/head
	Herbal preparations	Depending on product	

The occurrence of sheep scab and lice is also on the increase. Because of the severe welfare implications of scab infestation, organic farmers should aim for prevention and/or eradication by establishing closed flocks, bringing in stock only from scab-free sources and introducing quarantine and treatment of brought-in animals, measures which are also effective in preventing lice. Synthetic pyrethroid dips (e.g Flumethrin) have been withdrawn for environmental reasons. Organophosphates may only be used as a last resort for scab control and with permission by the control body (not permitted under Soil Association standards). They should be used and disposed of very carefully to avoid environmental pollution.

Internal parasites[21]

Control of intestinal (helminth) parasites is achieved primarily through preventive management practices, including clean grazing systems (alternate years), mixed stocking, rotational grazing, selection of resistant stock and appropriate stocking rates. Individuals or groups of animals and ewes at lambing showing evidence of parasite infestation may be drenched therapeutically, subject to withdrawal periods and other standards requirements. Industry guidance to minimize the spread of resistance should be followed (SCOPS – Sustainable Control Of Parasites in sheep, or COWS – Control Of Worms Sustainably). Husk (lungworm) in suckler calves may be controlled by development of natural immunity by grazing with adults. In dairy youngstock lungworm may be controlled by vaccination where a known farm problem exists. Dairy calves may be dosed with oral vaccine (Huskvac - available only on

---

21. *Controlling Roundworm in Organic Sheep*, Soil Association Technical Guide; free for Members.

prescription) before turn out. The husk vaccine is based on a natural challenge avoiding the risks of many other vaccines. Similarly, liver fluke should be controlled by strategic use of flukicides where there is a known farm problem and where fencing off risk areas on the farm is impractical.

Calculating livestock units for grazing animals

The Table shows the livestock unit equivalents for different types of livestock that can be used when calculating stocking rates.

Summary Grazing Livestock Units per head

Type	LU/head
Large, high-yielding dairy cow	1.6
Small, low-yielding dairy cow	1.0
Beef/suckler cow	0.75
Beef/dairy bull	0.65
Other cattle <1 year	0.34
1-2 years	0.65
>2 years	0.80
Beef/suckler cow with calf (to weaning at 8 months)	1
Finishing cattle (9 – 24 months)	0.74
Heifers per replacement unit (weaning – 27 months)	1.2 LU/head reared
Ewes with lambs to 6 months	0.11
Upland ewes with lambs to 6 months	0.08
Other sheep over 1 year (such as ram & ewe lamb)	0.08
Store lamb less 1 year	0.04

Reference: based on Nix Farm Management Pocketbook 2023 p.49 and own calculations. Different conversion factors may be used when calculating support payments.

Support payments for livestock and stock disposal

Most livestock payments have been replaced by direct payment schemes (see Section 5 for further details). Most regions of the UK and Ireland no longer offer support schemes specifically for livestock, although – while not specifically aimed at organic farms – some do exist in Scotland.

Scottish Suckler Beef Support Scheme

This scheme has replaced the previous Scottish Beef Calf Scheme. Payments are made to farms maintaining a suckler beef herd in Scotland, on calves that are at least 75% beef bred, were born in Scotland on or after 2 December 2014 and held there continuously for at least 30 days from birth, and that have not previously been supported under the Scottish Beef Scheme. Rates of payment are not fixed, depending on farm location and numbers of claimed animals. 2023 values estimated at £101 (mainland) and £144 (highland) per eligible animal.

https://tinyurl.com/Scot-suckler-beef

Scottish Upland Sheep Support Scheme

The Scottish Upland Sheep Support Scheme offers direct support to businesses reliant upon rough grazing. It is open to businesses with at least 80% of agricultural land in Scottish Basic Payment Region 3, with an LFA grazing category of A and no more than 200 ha of good quality agricultural land in Basic Payment Region 1. Payments are made on ewe hoggs born on eligible holdings that remain on site from 1st December in year claimed to 31 March the following year and are less than 12 months old at the start of the retention period. Payment rates are not fixed and will vary depending on the total number of animals claimed, depending on farm location and numbers claimed. Recent values are estimated to be in the region of £61 per eligible ewe hogg. https://tinyurl.com/Scot-upland-sheep

Further reading

Blair R (2017) *A Practical Guide to the Feeding of Organic Farm Animals.* 5M Publishing

Blair R (2021) *Nutrition and Feeding of Organic Cattle.* CABI Publishing

Viora L, Graham EM, Mellor DJ, Reynolds K, Simoes PB, Geraghty TE (2014) *Evaluation of a culture-based pathogen identification kit for bacterial causes of bovine mastitis.* Vet Rec. 26;175(4):89. doi: 10.1136/vr.102499. Epub 2014 Jul 10. PMID: 25013087.

Vaarst M. & Roderick S. (eds.), *Improving organic animal farming*, Burleigh Dodds Science Publishing, Cambridge, UK, 2019, (ISBN: 978 1 78676 180 4; www.bdspublishing.com)

Vaarst M. et al. (2004) *Animal Health and Welfare in Organic Agriculture*. CAB International: Wallingford.

Younie D (2012) *Grassland Management for Organic Farmers*. Crowood Press, Marlborough.

Websites, technical guides and leaflets

Frost D, Morgan M, Moakes S (2009) A farmers' guide to organic upland beef and sheep production. Technical Guide Organic Centre Wales.

SOLID (Sustainable Organic and Low Input Dairying) Technical Notes: Trace elements, Dairy Rations, Feeding Home grown Protein, Diverse swards, Dairy breeding, Suckling calves: https://agricology.co.uk/resource/sustainable-organic-and-low-input-dairying-solid/

Organic Farm knowledge: Section on animal health and welfare: https://organic-farmknowledge.org/discussion/theme/210

ICOPP (Improved Contribution of Local Feed to Support 100% Organic Feed Supply to Pigs and Poultry). At https://orgprints.org/id/eprint/28078/

ICOPP Technical Notes (2015) (No 1) 100% organic poultry diets: Concentrates,
(No 2) 100% organic poultry diets: Roughage and foraging from the range,
(No 3) 100% organic pig diets: Concentrates
(No 4) 100% organic pig diets: Feeding roughage and foraging from the range.

IOTA Technical Leaflets: Organic dairy cow nutrition (No 1) https://tinyurl.com/IOTA-dairy-nut & Organic beef & sheep nutrition (No3) https://tinyurl.com/IOTA-beef-sheep

IOTA Research Reviews. https://tinyurl.com/IOTA-poultry-nut

Burke G (2008). The role and management of herbal pastures for animal health, productivity and product quality. http://orgprints.org/13566;

Dinnage G (2008) Organic poultry nutrition and rations http://orgprints.org/5976;

Powell D (2008) The role and management of whole-crop forage for organic ruminants http://orgprints.org/13556/

Tame M (2008) Dairy Cow Nutrition http://orgprints.org/5982;

Tame M (2008) Management of trace elements and vitamins in organic ruminant livestock nutrition in the context of the whole farm system http://orgprints.org/13565

Nixey C, Little T (2013) Making poultry feed more sustainable: Dehulling homegrown protein crops. Report for Organic Centre Wales.

Agricology factsheet. Sustainable options for producing home-grown organic poultry feed. At: https://tinyurl.com/agricology-poultryfeed

Roderick S (2008) Dairy cow breeding for organic farming. IOTA Research review. htpp://orgprints.org/5975.

Soil Association free technical guides: Organic sheep production; organic pig production; organic poultry production – an introduction.
Factsheets: organic withdrawal periods; restricted antibiotics; fecal Egg Count.
Available at https://www.soilassociation.org/farmers-growers/low-input-farming-advice/technical-guides/.
Other technical guides free for members only: Managing Internal Parasites in organic cattle; Organic Veal Production; Grassland management.

Dairy cows – Friesian/Holstein all year round calving

Output[1]

Milk yield Total milk yield includes milk for calf feeding as transfer to dairy young stock enterprise, valued at market price. Milk yields vary widely from 5,000-8,000 litres per cow. Average of Kingshay Dairy Manager[2] benchmarked farms 6,600 litres and a herd size of 236 cows.

Price and organic premiums Currently, average annual organic milk price paid by major buyers is 40 - 51 p/l for level supply of around 1 million litres (net price after testing, transport and other co-op costs, excluding membership fees). Some buyers offer quality and/or seasonality pricing. Farm gate price increases may be offset by increases to feed and energy prices. Conventional price of milk is expected to drop throughout 2023, but it is unclear if organic milk will track conventional. Arla reported a price drop to under 40.88p/l May 2023 (down from 51p/l at the end of January). Hygiene Band A assumed. Channel Island cow yields are typically about 1,500 litre lower (Guernseys typically yield 250 litres higher than Jerseys) but the specialist demand is low and very few buyers are willing to reward the quality with a higher price.

Milk from forage: Average value ranges from 2-4,000 l per cow, average 3,300 l

Margins over purchased feed and feed and fertiliser refer to milk sales only.

Mortality/replacement rates may be lower in established organic herds.

Calf sales Prices are estimates for standard calf sales with assumed split of 40:60 for dairy and beef crosses. With the current high rearing costs to beef price ratio there is limited potential for rearing dairy calves for finishing.

Variable costs

Concentrates Compound feed with 100% organic components assumed; in-conversion max 30% of dry-matter intake (60% if from own holding). For price range and nutritional value for some feeds see p.185.

Bulk feeds Illustrated using a 10% organic premium over conventional prices.

Vet and med Costs vary widely (£82-94/head), lower in established organic herds. Average £82/cow. Assumed costs exclude £5-10 per cow for animal health plans.

Other includes consultancy, consumables and dairy stores.

1. Hulsen, J Cow Signals–A practical guide for dairy farm management www.roodbont.nl; Obsalim cards to help with daily diagnosis of nutrition: www.obsalim.com/en.
2. www.kingshay.com/dairy-costings.
3. https://tinyurl.com/Arla-May23

Bedding — Straw costs have not been included – price £50-85/t, depending on location (300-500kg/cow required for cubicle housing, 1.5-2 t/cow in straw yards, for 200-day housing period). For costs for manure/ slurry handling see Section 14.

Forage — Based on 4-year ley and costs for 8.5t silage per cow and on 4-year ley grazing and permanent grassland plus some silage for replacement.

Rearing for dairy replacements (see p198) — Whole milk assumed, valued at the average organic price. May be fed from suckler cow or bucket. Restricted bucket feeding assumed (350 litres) with weaning at 12 weeks. Higher milk requirements (400-450 litres) with unrestricted feeding and extended weaning period (up to 16 weeks).

Dairy cows — *Friesian/Holstein - all year round calving*

Physical assumptions

Calving index	380 days	6.0% calf mortality	1.3% cow mortality
Stocking rate	0.67 ha/cow	1.5 cows/ha	1 LU/cow
Milk quality	4.1% butter fat	TBC	20 Band A
	3.35% protein		
Concentrate use	0.26 kg/litre		

Financial data				*£/cow*		*p/litre*
Milk sales /tranfers	6500 litres/cow	@	43.0 p/litre	2795		43.0
Calves	0.90	@	120 £/head	108		1.7
Cull cows	20.7%	@	1146 £/head	237		3.6
Replacement heifers	22%	@	1785 £/head	-393		-6.04
Total output					**2748**	42.3
Concentrates	1690 kg/cow	@	585 £/t	989		15.2
Purchased bulk feed	200 kg/cow	@	90 £/t	18		0.3
Vet & med (incl. health plan)				90		1.4
AI and recording				60		0.9
Other				70		1.1
Total variable costs					**1227**	18.9
Margin over purchased feed					1788	27.5
Gross margin excluding forage					**1521**	23.4

	per ha *(per ac)*	per cow	*p/litre*
Forage costs (£)	143 *(57)*	**96**	1.5
Milk from forage (litres)	4853 *(1941)*	3235	
Margin over feed and fertilisers (£)	2590 *(1036)*	1726	26.6
Gross margin including forage (£)	2139 *(855)*	**1426**	21.9

Sensitivity analysis — For explanation, see p. 104

	Change in value (+/-)	*Change in margins (+/-)* £/cow	p/l	£/ha *(£/ac)*	*Value range* Low	High	*GM ex. forage range (£/cow)*	
Milk yield	100 litres	28	0.07	42 *(17)*	4700	8000	1021	1938
Milk price organic	1 p/litre	65	1.00	98 *(39)*	40.0	55.0	1326	2301
Milk price conventional	-1.5 p/litre	-97.5	-1.50	-146 *-(59)*	35.0	46.0	1001	1716
Concentrate price	10 £/t	-17	-0.26	-25 *-(10)*	530	630	1445	1614
Concentrate use	0.01 kg/litre	-38	-0.59	-57 *-(23)*	0.15	0.30	1369	1939

	Change in value (+/-)	*GM change* £/ha *(£/ac)*	*Value range* Low	High	*Gross margin (inc. forage) range* £/ha *(£/ac)*	£/ha *(£/ac)*
Stocking rate	0.1 cow/ha	152 *(61)*	1.2	1.7	1682 *(673)*	2443 *(977)*

Dairy replacements *Friesian/Holstein*

Physical assumptions

Calving time	All year	Age at calving	27 months	Mortality	5%
Stocking rate	1.2 LU/replacement unit (calf & yearling)			1.50 LU/ha	1.26 RU/ha

Financial data

				£/replacement		
Replacement heifer	1785 £/heifer			1785		
Less calf cost	1.05 calves	@	210 £/head	-221		
Total output					**1564**	
Whole milk	350 litres	@	43 p/litre	151		
Concentrates 0-12 wks	75 kg	@	420	32		
Concentrates 12wks - 24mtl	350 kg	@	380 £/t	133		
Purchased bulk feeds	200 kg	@	90 £/t	18		
Vet. and med.				25		
AI and other livestock expenses				40		
Total variable costs					**398**	
Gross margin excluding forage					**1166**	
						£/ha *(£/ac)*
Forage costs					**100**	126 *(50)*
Gross margin including forage					**1066**	1344 *(538)*

Sensitivity analysis

For explanation, see p. 104

	Change in value (+/-)	*Change in GM (£/RU)*	*Value range Low*	*High*	*Gross margin range excluding forage (£/RU)*	
Heifer price	10 £/head	10	1300	1600	681	981
Milk price	1 p/l	-3.5	40.0	55.0	1124	1176

SECTION 11
LIVESTOCK GROSS MARGINS

Dairy cows – Friesian/Holstein, spring calving

Output[1]

Milk yield	Start of calving needs to ensure that peak milk production is achieved at peak grass growth. A tight calving pattern will significantly affect the supply profile and milk price. Milk buyer requirement and pricing are very important and must be considered.
Price and organic premiums	Assumed lower than in all year round calving because of seasonality. Milk buyers' pricing will affect milk price significantly especially for Channel Island and crossbreed cow.

Variable costs

Forage	Based on 40% permanent pasture, 4-year ley and 7.25t/silage per cow.
Other costs	Substantial savings in overhead costs (particularly in terms of labour and young stock rearing costs) can be achieved by having a tight calving pattern.

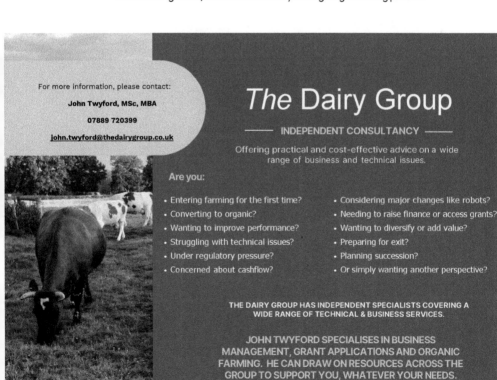

Dairy cows	*Friesian/Holstein - spring calving*

Physical assumptions

Calving index	375 days	6.0% calf mortality	2.0% cow mortality	
Stocking rate	0.63 ha/cow	1.6 cows/ha	0.9 LU/cow	
Milk quality	4.25% butter fat			
	3.25% protein	TBC	20 Band	A
Concentrate use	0.15 kg/litre			

Financial data

				£/cow		*p/litre*
Milk sales/transfers	5,000 litres/cow	@	41.0 p/litre	2050		41.0
Calves (adjusted for mortality and calving index)		@	120 £/head	108		2.2
Cull cows	18.0%	@	1146 £/head	206		4.1
Replacement heifers	20%	@	1785 £/head	-357		-7.1
Total output					**2008**	40.2
Concentrates	500 kg/cow	@	585 £/t	293		5.9
Vet & med (incl. health plan)				50		1.0
AI and recording				60		1.2
Other				70		1.4
Total variable costs					**473**	9.5
Margin over purchased feed					1758	35.2
Gross margin excluding forage					**1535**	30.7

	per ha *(per ac)*	per cow	*p/litre*
Forage costs (£)	126 *(50)*	**79**	1.6
Milk from forage (litres)	6455 *(2582)*	4034	
Margin over feed and fertilisers (£)	2713 *(1085)*	1696	33.9
Gross margin including forage (£)	2331 *(932)*	**1457**	29.1

Sensitivity analysis

For explanation, see p. 104

	Change in value (+/-)	*Change in margins (+/-)* £/cow	p/l	£/ha *(£/ac)*	*Value range Low*	*High*	*GM ex. forage range (£/cow)*	
Milk yield	100 litres	32	0.03	52 *(21)*	3200	6000	955	1857
Milk price	1 p/litre	50	1.00	80 *(32)*	38.0	53.0	1385	2135
Milk price conventional	-1.5 p/litre	-75	-1.50	-120 -*(48)*	35.0	46.0	1235	1785
Concentrate price	10 £/t	-5	-0.10	-8 -*(3)*	530	630	1513	1563
Concentrate use	0.01 kg/litre	-29	-0.59	-47 -*(19)*	0.10	0.26	1213	1681

	Change in value (+/-)	*GM change* £/ha *(£/ac)*	*Value range* *Low*	*High*	*Gross margin (inc. forage) range* £/ha *(£/ac)*	£/ha *(£/ac)*
Stocking rate	0.1 cow/ha	154 *(61)*	1.3	1.8	1870 *(748)*	2638 *(1055)*

SECTION 11 LIVESTOCK GROSS MARGINS

Beef production

Output

Calving	Spring calving with single suckling assumed for suckler cows.
Price	Assumed for finished animals, based on typical organic prices for last season. Organic store cattle prices are related to organic finished cattle. Price penalties below 270kg dcw of 20p/kg.
Organic premium	The organic market is relatively stable and a small organic premium is available to most producers if carcass quality meets requirements and supply is programmed with buyers.
Finishing weight	Varies according to age, breed and sex. Mixed sex, beef crosses assumed, e.g., Charolais X: 290-340kg dcw; Hereford X: 260-300kg dcw. Market specifications are max 380kg, min 260-270kg (490-510kg lw) depending on stock type and outlet.
Replacements	Suckler herds are assumed to be self-contained with heifers calving at 3 years, and a proportion of calves for replacements retained each year. No value for purchasing replacement heifers has been included. Cull cow price at £4/kg dcw.
Bulls	1 for 35 cows, 4 years, replacement value £3,000, cull value £1,200.
Purchased calves	Organic status calves purchased to replace losses.

Variable costs

Concentrates	Based on 40% permanent pasture, 4-year ley and 7.25t/silage per cow.
Bulk feeds	Hay purchased at organic prices.
Bedding	Straw costs have not been included (price £50-85/t), depending on location. Usage levels variable, typically 1-2t/LU for 200-day winter in straw yards. Non-organic straw is permitted for bedding.
Forage	These are based on the net annual variable costs for 4-year ley and permanent grassland. Hill/upland farms will have lower forage costs due to higher proportion of permanent pasture and rough grazing (typically £15-30/ha).
Replacements	Variable costs for suckler cows include allowance for replacement rearing.
Transport and miscellaneous	Including levies, commission, tags, recording.

Suckler cow - Lowland with finishing at 24 months

Physical assumptions

				Livestock units
Suckler cows	7	years in herd	1% mortality	0.75 LU/cow
Bulls	4	years in herd	35 cows/bull	0.65 LU/bull
Calves	0.94	born/cow/year		0.03 LU/calf
Progeny to 24 mths	0.9	per cow		0.74 LU/finished animal
Replacement bulling heifers	0.15	retained per cow		
Heifers sold finished	0.3	per cow	270 kg dcw/head	
Steers sold finished	0.45	per cow	324 kg dcw/head	
Age at finish	24	months		
Liveweight, average	560	kg per head	54% killing out	
Stocking rate	1.1	LU/ha	0.75 cows/ha	1.5 LU/cow

Financial data

				£/cow	
Finished cattle sales	0.75 head/c	302 kg dcw	5.35 £/kg dcv	1213	
Cull cows	0.14 /cow	@	1200 £/head	168	
Bull depreciation	0.0071 /cow	@	-3000 £/bull	-21	
Purchased calves	0.05 /cow	@	150 £/head	-8	
Total output					**1352**
Concentrates: cows	200 kg	@	540 £/t	108	
Concentrates: beef	270 kg	@	405 £/t	109	
Purchased bulk feed	250 kg	@	90 £/t	22.5	
Vet. and med.				42	
Transport and misc.				30	
Total variable costs					**312**

		£/ha *(£/ac)*
Gross margin per cow excluding forage	**1041**	
Forage costs per cow and progeny	**148**	112 *(45)*
Gross margin per cow including forage	**892**	672 *(269)*

Sensitivity analysis

For explanation, see p. 104

	Change in value (+/-)	*Change in GM (£/cow)*	*Value range Low*	*High*	*Gross margin range excl. forage (£/cow)*	
Sale price	0.1 £/kg dcw	30.24	5.1	5.5	965	1086
Sale conventional	-0.1 £/kg dcw	-30.2	4.4	5	753	935
Sale weight	10 kg dcw	53.5	260	350	814	1295
Concentrate use	10 kg	5.4	350	550	1084	976
Concentrate price	10 £/t	-4.7	360	615	1005	1125
Calving percentage	0.01 %	16.18	0.8	0.98	814	1105

Suckler cows - Lowland sold as stores

Physical assumptions

					Livestock units	
Suckler cows	7	years in herd	1%	mortality	0.75	LU/cow
Bulls	4	years in herd	35	cows/bull	0.65	LU/bull
Calves	0.94	born/cow/year	5%	mortality		
Calf sale/transfer at	8.3	months	265	kg lw	0.24	LU/calf
Replacements	0.15	calves retained per cow			0.23	LU/cow
Stocking rate (incl repl)	1.10	LU/ha	0.90	cows/ha	1.22	LU/cow
						incl. replacement & calf to 250 days

Financial data

						£/cow	
Calf sales	0.79	/cow	265 kg lw	2.50	£/kglw	521	
Cull cows	0.14	/cow	@	1200	£/head	171	
Bull depreciation	0.0073	/cow	@	-3000	£/bull	-22	
Purchased calves	0.05	/cow	@	150	£/head	-8	
Total output							**664**
Concentrates	200	kg	@	405	£/t	81	
Purchased bulk feed	150	kg	@	90	£/t	14	
Vet. and med.						35	
Transport and misc.						30	
Total variable costs							**160**
Gross margin excluding forage							**504**

	£/cow	£/ha	*(£/ac)*
Forage costs	**124**	112	*(45)*
Gross margin including forage	**380**	343	*(137)*

Sensitivity analysis

For explanation, see p. 104

	Change in value (+/-)	*Change in GM (£/cow)*	*Value range Low*	*High*	*Gross margin range excl. forage (£/cow)*	
Sale/transfer price	0.10 £/kg lw	27	2.10	2.90	398	610
Sale conventional	-0.71 £/kg lw	-188	1.70	2.40	292	478
Sale/transfer weight	1.0 kg lw	2.5	280	310	542	617
Concentrate use	10 kg	4.05	100	250	524	464
Concentrate price	10 £/t	-2.00	380	460	493	509
Calving percentage	1%	6.63	85%	90%	444	478

Beef finishing - spring-born stores, finished at 22-26 months

Physical assumptions

Age at transfer/finish.	9 months to	24 months =	458 days
Liveweight	265 kg purchased	550 kg at sale	0.62 kg daily lw gain
Deadweight	54% killing out	297 kg dcw	
Animals	50% male	50% female	4% mortality
Stocking rate	0.74 LU/finished hd	1.10 LU/ha	1.50 head/ha

Financial data

					£/head		
Finished animal	297 kg dcw		@	5.35 £/kg	1589		
Less store animals	265 kg lw	104%	head	2.50 £/kg	-689		
Total output						**900**	
Concentrates	270 kg		@	540 £/t	146		
Purchased bulk feeds	100 kg		@	90 £/t	9		
Vet. and med.					18		
Transport and misc.					30		
Total variable costs						**203**	
Gross margin excluding forage						**697**	
							£/ha *(£/ac)*
Forage costs						**75**	112 *(45)*
Gross margin including forage						**622**	932 *(373)*

Sensitivity analysis

For explanation, see p. 104

	Change in value (+/-)	*Change in GM (£/head)*	*Value range Low*	*High*	*Gross margin range excluding forage (£/hd)*	
Sale price	0.10 £/kg dcw	30	4.60	5.50	474	742
Sales price, conventional	-0.60 £/kg dcw	-162	4.10	5.00	360	603
Sale weight	10 kg dcw	53.5	270	350	553	981
Concentrate use	10 kg	-5	200	500	573	735
Concentrate price	10 £/t	-3	290	560	692	765

Sheep production

Output

Price	The organic lamb market is currently oversupplied at some times of the year, hence the premium is relatively small and may not be available for all lambs, particularly those that do not meet specification. The prices assumed for finished animals (R3L-carcass) are typical for organic prices 2022-23. Prices vary throughout the season; continuity of supplies throughout the year remains a problem. Store values assumed at £65/head. Store producers should establish links with finishers, but the organic store market is small. Organic premium specification is generally 14.5 to 23 kg dcw. Deduction for lamb between 13 and 14.5kg dcw is 20p/kg; under 13kg is 40p/kg. There is a deduction of 100p/kg for every kg over 23kg. The value of the premium price may be significantly eroded by additional haulage and marketing costs. The lower sales price in the sensitivity analysis for upland ewes reflects the impact of selling at conventional prices.
Finishing weight	Varies according to breed. Breeds suited to supplying supermarkets and similar outlets assumed for lowland systems; lighter breeds assumed for uplands although less suitable for organic markets.
Lambing date	March/April lambing has been assumed. Delaying lambing may be advantageous in organic systems, allowing outdoor lambing with reduced disease and Nematodirus parasite risk, labour requirements, bedding, and feed costs, as well as finishing lambs during the higher price periods in mid-late winter. Early lambing for the spring market is also an option, but higher concentrate feeding is required.
Replacements	The sheep flock is assumed to be self-contained, which is important for disease management in organic systems.

Variable costs

Concentrates	Proprietary concentrates assumed. Whole or rolled cereals may be used, if forage quality good enough, reducing the cost.
Bulk feeds	Hay purchased at organic prices.
Bedding	Straw costs have not been included. Av. price £50 - 85/t, depending on location. Usage levels variable, typically 100 kg/ewe for 100 days housed.
Forage	These are based on net annual variable costs for 5-year leys and permanent pasture and permanent pasture only for upland farms. Some hill/upland farms have lower forage costs due to higher proportion of permanent pasture and rough grazing (typically £15-30/ha).
Replacements	Variable costs for ewes include allowance for replacement rearing.
Transport and miscellaneous	Including levies, shearing, commission, tags, recording.

Lowland breeding ewes

Physical assumptions					*Livestock units*
Ewes	5 years in flock		3% mortality		0.08 LU/ewe
Rams	3.0 years in flock		40 ewes/ram		0.08 LU/ram
Lambs per ewe	1.60 born	8% mort.	1.47 reared		
Lamb allocation	80% finishec	4% stores	16% retained		
Lamb sale/transfer at	6 months	38 kglw	46% killing out		0.04 LU/lamb
Replacements per ewe	0.2 reared	3% mort.	0.2 replacements		0.16 LU/replacement
Stocking rate	1.00 LU/ha		5.9 ewes etc./ha		0.17 LU/ewe

Financial data				*£/ewe*		
Finished lambs	1.18 /ewe	@	5.20 £/kg dcv	107.0		
Store lambs	0.06 /ewe	@	65 £/head	3.7		
Culled stock	0.21 /ewe	@	90 £/head	18.8		
Wool	2 kg/ewe	@	1.00 £/kg	2.0		
Replacement rams	0.01 /ewe	@	500 £/head	-4.2		
Total output					**127.3**	
Minerals	1.8 kg/ewe	@	510 £/t	0.9		
Concentrate	40 kg/ewe	@	405 £/t	16.2		
Purchased bulk feeds	15 kg/ewe	@	90 £/t	1.4		
Vet. and med.				8.0		
Transport and misc.				14.0		
Total variable costs					**40.5**	
Gross margin excluding forage					**86.9**	
						£/ha *(£/ac)*
Forage costs					**16.6**	99 *(40)*
Gross margin including forage					**70.2**	418 *(167)*

Sensitivity analysis For explanation, see p. 104

	Change in value (+/-)	*Change in GM (£/ewe)*	*Value range Low*	*High*	*Gross margin range excl forage (£/ewe)*	
Sale price	0.10 £/kg dcw	2.1	4.90	6.90	80.7	121.9
Sale price, conventional	-0.40 £/kg dcw	-8.2	4.80	6.50	78.6	113.6
Sale weight (finished)	1.0 kglw	2.82	30	45	64.3	106.6
Lambs finished/ewe	0.1 /ewe	9.1	1.00	1.80	70.7	143.5

Upland breeding ewes

Physical assumptions

				Livestock units
Ewes	4 years in flock		5% mortality	0.06 LU/ewe
Rams	3.0 years in flock		40 ewes/ram	0.08 LU/ram
Lambs per ewe	1.40 born	10% mort.	1.26 reared	
Lamb allocation	65% finished	11% stores	24% retained	
Lamb sale/transfer at	6 months	35 kglw	46% killing out	0.04 LU/lamb
Replacements per ewe	0.31 reared	4% mort.	0.30 replacements	0.12 LU/replacement
Stocking rate	0.80 LU/ha		5.9 ewes etc./ha	0.14 LU/ewe

Financial data

				£/ewe	
Finished lambs	0.82 /ewe	@	5.20 £/kg dcw	68.6	
Store lambs	0.14 /ewe	@	60 £/head	8.5	
Culled stock	0.26 /ewe	@	70 £/head	18.1	
Wool	1.3 kg/ewe	@	0.35 £/kg	0.5	
Replacement rams	0.01 /ewe	@	500 £/head	-4.2	
Total output					**91.4**
Minerals	1.8 kg/ewe	@	510 £/t	0.9	
Concentrate	40 kg/ewe	@	405 £/t	16.2	
Purchased bulk feeds	20 kg/ewe	@	90 £/t	1.8	
Vet. and med.				8.0	
Transport and misc.				14.0	
Total variable costs					**40.9**

		£/ha *(£/ac)*
Gross margin excluding forage	**50.5**	
Forage costs	**12.2**	72 *(29)*
Gross margin including forage	**38.3**	226 *(90)*

Sensitivity analysis

For explanation, see p. 104

	Change in value (+/-)	*Change in GM (£/ewe)*	*Value range Low*	*High*	*Gross margin range excl forage (£/ewe)*	
Sale price	0.10 £/kg dcw	2.9	4.90	6.90	41.9	99.2
Sale price, conventional	-0.40 £/kg dcw	-11.5	4.80	6.50	39.0	87.7
Sale weight (finished)	1.0 kglw	1.96	30	40	40.7	60.3
Lambs sold/ewe	0.1 /ewe	8.4	0.80	1.40	48.9	99.1

Pig production[4]

Output

Price Premiums are essential to cover costs of production, irrespective of whether producing for multiple retailers or for local markets (see Section 2). Sensitivity analysis for impact of production without premium. Premium prices may be affected by additional haulage and marketing costs (approx 90p-100p per pig). Penalties on price will be based on fat depth (max. 13mm back fat) and lean to fat ratios and may apply for animals over target weight of 80kg. Significantly higher prices may be obtained through farm shop or retail sales. Killing charges of £36-70/pig depending on weight, cutting for pork £0.60/kg, sausage making and ham curing £3.50/kg.

Weaning Assumed at 8 weeks allowing 1.9 litters per year, 20 weaners per sow and year. Prices depend on feed price and finished sales value.

Finishing Weight varies according to end purpose, baconers assumed with porkers indicated in the sensitivity analysis. Standards require the possibility to root and dung in exercise area. Finishing times to reach required weight varies depending on breed and season with shorter times in warmer weather.

Sows Breed needs to be suited to free-range systems, e.g., Saddleback (these breeds tend to be less efficient in terms of feed conversion than Landrace or Large White). May be crossed with Duroc, Landrace or other sires for better carcass quality. Organic hybrid gilts are now available.

Boars Enter herd at 3-4 months old.

Replacements The herd is assumed to be self-contained, raising its own gilts. Gilts served at 8-9 months old (100–130kg lw).

Site Outdoor pigs need a low rainfall area and free draining site. Typically rotated every 12 weeks on last year of ley before arable phase of rotation, avoiding the need for nose ringing, which is not allowed under organic standards. Rotation of site (1-2 times per year) is essential to avoid build-up of parasitic disease such as Ascaris spp. Stocking rate will vary depending on the availability of land, site suitability. Typical range is 2.5–6 sows/ha for outside finishing; indoor finishing is no longer permitted by Article 95 (3) of 889/2008.

Variable costs

Feed Prices are based on bulk delivery of compound feed. Feed prices vary through the year and with type of delivery. Requirements can be reduced by more forage[5].

Bedding Straw costs have not been included, priced at £50 - 85/t. Usage levels are typically 300-330kg/litter if finished outdoors.

Forage These are based on net annual variable costs from medium/long term leys (p 170).

Replacements	Variable costs for sows include allowance for replacement rearing.
Housing	Not included.
Fencing and water troughs	Not included. Assume 2 sows/0.1ha (0.25ac) farrowing paddock, 8-10 dry sows/0.2ha (0.5ac) paddock, 30-40 finishing pigs per 0.2-0.4ha (0.5-1ac) paddock. If all pigs kept outdoors then 80–100m2/sow incl. finishing groups.

Pigs (outdoor breeding)

Physical assumptions							Livestock units	
Sows	3	years in herd	5%	mortality			0.44	LU/sow & litters
Boars	2.5	years in herd	10	sows/boar			0.35	LU/boar
Litters per year	1.9	with	10 piglets	15% mortality				
Weaners per sow	8.5	reared/litter	16	reared/year			0.09	LU/finisher
Weaner sale/transfer at	12	weeks	28 kglw	2.8% post weaning mortality				
Replacement gilts	0.42	/sow					0.17	LU/replacement
Stocking rate	5.00	LU/ha	2.52	sows etc. /ha			1.99	LU/sow etc.

Financial data						£/sow		
Weaners (excl. gilts)	15.13	/sow	@	4.20	£/kg lw	1780		
Culled stock	0.37	/sow	@	100	£/head	37		
Replacement boars	0.04	/sow	@	1250	£/head	-50		
Total output							1767	
Boars and gilts feed	1000	kg	@	600	£/t	600		
Lactation feed	600	kg	@	650	£/t	390		
Weaner feed	45	kg	@	680	£/t	31		
Vet. and med.	3.00	£/weaner				78		
Miscellaneous	2.00	£/weaner				30		
Total variable costs							1129	
Gross margin excluding forage							638	
								£/ha *(£/ac)*
Forage costs							51	128 *(51)*
Gross margin including forage							587	1478 *(591)*

Sensitivity analysis For explanation, see p. 104

	Change in value (+/-)		*Change in GM (£/sow)*	*Value range Low*	*High*	*Gross margin range excluding forage (£/sow)*	
Sale price	0.01	£/kg lw	4	3.50	5.00	342	977
Sale price, conventional	-0.06	£/kg lw	-25	1.50	2.00	-506	-294
Weaners reared/sow	1	/sow	35	14	17	569	673
Sow feed use	100	kg/sow	62	1400	1800	514	762
Concentrate price	10	£/t	16	550	700	474	720

4. *The Manual for Eastbrook Farm Organic Pigs*. From: victoria@helenbrowningorganics.co.uk
5. *Pig Ignorant? A Soil Association Technical Guide to Small-Scale Pig Keeping*. Bristol.

Pigs (outdoor finishing - baconers)

Physical assumptions

Age at transfer/finish	12 weeks to	26 weeks	98 days
Liveweight	28 kg at transfer	110 kg at sale	0.84 kg daily lw gain
Deadweight	72 % killing out	79 kg dcw	2% mortality
Feed conversion	2.45 kg /day	240 kg feed	2.93 kg feed/kg lwg
Stocking rate	forage area included in breeding enterprise		

Financial data

				£/pig finished	
Finished pigs	79 kg dcw	@	3.75 £/kgdw	297	
Less weaners	1.02 weaners	@	118 £/head	-120	
Total output					**177**
Concentrates per pig	240 kg	@	650 £/t	156	
Vet. and med.				2	
Transport and misc.				9	
Total variable costs					**167**
Gross margin excluding forage					**10**

Sensitivity analysis

For explanation, see p. 104

	Change in value (+/-)	*Change in GM (£/pig)*	*Value range Low*	*High*	*Gross margin range excluding forage (£/pig)*	
Sale price	0.1 £/kg dcw	7.92	3.50	4.25	-10	50
Sale price conventional	-1.0 £/kg dcw	-79.2	1.70	2.50	-152	-89
Sale weight	1 kg dcw	3.75	60	85	-62	32
Concentrate use	10 kg/pig	-6.50	220	280	-16	23
Concentrate price	10 £/t	-2.40	600	700	-2	22

Porkers

Age at transfer/finish	12 weeks to	19 weeks	49 days
Liveweight	32 kg purchased	70 kg at sale	0.78 kg daily lw gain
Deadweight	70 % killing out	49 kg dcw	2% mortality
Feed conversion	2 kg/day	98 kg feed	2.58 kg feed/kg lwg
Finished pigs	49 kg dcw	@ 3.75 £/kgdw	184 £/pig
Concentrate costs	64 £/pig	Gross margin	-11 £/pig

Management notes

Site Outdoor pigs need a low rainfall area and free draining site. Typically rotated every 12 weeks on last year of ley before arable phase of rotation, avoiding the need for nose ringing, which is not allowed under organic standards. Rotation of site (1-2 times per year) is essential to avoid build-up of parasitic disease such as Ascaris spp. Stocking rate will vary depending on the availability of land, site suitability. Typical range is 2.5–6 sows/ha for outside finishing; indoor finishing without exercise area is no longer permitted by Article 95 (3) of 889/2008. Standards require the possibility to root and dung in exercise area.

Laying hens (producer-packer)

Output[6]

Price The market for organic eggs is growing slowly mainly reliant on UK supplies. Wholesale prices are assumed. Prices vary depending on which UK organic standard eggs are produced to. Producing for pack houses to EU and Soil Association standard is illustrated in the sensitivity analysis.

Variable costs

Pullets[7] Full organic rearing assumed. Smaller numbers of fully organically reared pullets are available, but larger numbers should be contracted to ensure supply. Full organic birds are likely to have advantage in terms of being used to the range; bought in bulk are likely to be priced from £9.50 upwards, depending on age.

With uncertainty over the future development there has been limited uptake of commercial organic rearing. Since the demand for commercially scaled organic pullets exceeds the available supply from organic rearing sources, Defra has continued a derogation allowing the use of conventional pullets until 31 December 2025[8]. Non-organic pullets bought for organic egg production have to be younger than 18 weeks and should only be chosen if no organic alternative is available. Some producers have set up their own on farm rearing facilities or have their organic pullets supplied by packers. Currently commercially available conventional pullets are in the region of £7.25-£10/pullet. Within the current legislation, pullets can be reared to organic standards with respect to feed and veterinary standards from day old and subsequently reared to full organic standard.

Feed Organic feed is preferred. Until 31 December 2025 a 5% proportion of non-organic feed is allowed. Fishmeal may be used, and diets may be supplemented with enzymes and microorganisms authorised under EU Directive 70/524/EE, minerals and vitamins (natural or synthetic identical to natural). The use of synthetic amino acids and coccidiostats is not permitted.

Additional costs Labour for egg collection and flock monitoring as well as time needed to market eggs.

6. Egg Grading and Packing OF&G Technical Guide TL211 www.organicfarmers.org.uk
7. Little, T et al. Developing the Supply of organic Pullets in Wales 2011 www.organiccentrewales.org.uk/uploads/pullet_factsheet_final.pdf
8. www.legislation.gov.uk/eur/2008/889/article/42

Management notes

Standards Some Control Bodies operate stricter standards than the retained EU Regulations.

Stocking rate The outdoor stocking rate should not go beyond 2,500 birds per hectare (equivalent to 4m² per bird) for pullets. Lower stocking rate of 10m²/bird (or 1,000 birds/ha) required under Soil Association and OF&G standards are assumed for adult layers. Both stocking rates must not exceed the limit of 170kg N/ha over the whole organic area. Higher stocking rates than 170kg N/ha are possible if an agreement is made with another organic holding to utilise the excess manures produced and the combined nitrogen rates of both holdings meets the requirements of the regulation. Provision of overhead shelter on range is required.

Housing Mobile or static housing possible, ideally well insulated to assist temperature regulation. The Soil Association tolerates a stocking density of 3,000 birds per house for laying hens and 1,000 birds per house for any other poultry. Floor area: 6 hens/m²; perch: 18 cm perch per bird; nest space: 7 hens/nest (6 for SA) or 120cm²/bird. Initial capital costs for mobile house £67/bird for 200-300 hens, depreciated over 10 years. Static house for 1,000-3,000 birds, initial capital cost at £54-66/bird complete, depreciate over 10 years.

Health Parasite and disease problems, including coccidiosis, salmonella and range parasites should be controlled by management, resting periods and vaccination in the first instance. Only in exceptional circumstances should conventional treatment be necessary. There are a number of acceptable products for the control of red mite. The National Control Programme for the control of salmonella in laying flocks came into force 1st February 2008 making monitoring at the following dates mandatory: pre-housing, 24-28wk plus continuous 15 wk.

Welfare Most significant problem is feather-pecking and cannibalism. Routine beak-trimming is not permitted by EC and SA Regulations. A preventive strategy using a wide range of options to avoid this problem is necessary.

Laying hens (producer-packer)

Physical assumptions

Laying period	50 weeks	8% mortality	1000 hens per ha
Eggs	280 eggs/year	6.6% seconds	4.5% smalls

Financial data

					£/hen/year	
Eggs	First Class	249 eggs/year	@	4.20 £/doz	87.12	
	Second class	18 eggs/year	@	2.40 £/doz	3.70	
	Smalls	13 eggs/year	@	1.80 £/doz	1.89	
Less pullet		1.08 birds/year	@	12.15 £/bird	-13.12	
Total output						**79.59**
Feed		50 kg/year	@	740 £/t	37.00	
Packaging		41 boxes	@	14 p/box	5.81	
Grading/packing		23 doz	@	15 p/doz	3.50	
Other livestock expenses					1.65	
Total variable costs						**47.96**
Gross margin excluding forage						**31.63**

Additional costs		£/h	£/hen	GM
Labour (welfare & egg collection)	1.5 days/1000 hens @	14.00	7.67	23.96
(marketing)	10 p/dozen		2.07	21.89

Sensitivity analysis

For explanation, see p. 104

	Change in value (+/-)	*Change in GM (£/hen)*	*Value range Low*	*High*	*Gross margin range excluding forage (£/hen)*	
Sale price	0.1 £/doz.	2.07	1.15	4.50	-31.6	37.9
Egg production	10 /hen	3.27	250	300	36.0	253.3
Concentrate use	1 kg/hen	0.74	40	60	24.2	39.0
Concentrate price	10 £/t	0.50	350	750	24.2	320.2
Pullet price	0.1 £/pullet	0.11	9.0	10.5	28.2	29.8

Sold to packer

				£/hen Output	*Gross margin*
Eggs (EU standards)	280 eggs/year	@	1.45 £/doz	20.71	-17.94
Eggs (Soil Association standard)	280 eggs/year	@	1.65 £/doz	25.38	-13.27

Table poultry (producer-killed and dressed)[9]

Output

Price Fresh poultry sales in multiple retailers are growing slowly. Prices shown represent direct sales to consumers.

Variable costs

Chicks Breed choice is an issue because of welfare problems associated with modern hybrids, more suitable slower maturing strains are becoming available. Modern breeds are used with appropriate management practices to minimise problems. If organic chicks are not available, non-organic chicks may be used until 31st of December 2025, but but they must be managed to organic standards from 3 days old.

Feed[10] For regulation requirements see layers; bulk purchase assumed. Feed requirements higher than conventional because of longer finishing periods but costs can be reduced by greater reliance on home grown grain. Vermin can cause considerable feed losses, especially in mobile/outdoor units. Costs are even higher in winter, especially if finishing in winter with poorly insulated housing. For production under contract no feed costs are assumed. Possible 100% organic feed rules coming in are likely to lead to increases in feed prices but the 5% conventional feed rule will remain effective until December 2025.

Range Access to range for at least 1/3 of lifetime. Minimum area 4m²/bird (2,500 birds ha), not exceeding 170kgN/ha.

Housing Costs not included. Max. group size of 4,800 chicken but lower with some Control Bodies. Max. 10 birds/m² to a max 21kg/m2 (EU/SA-Reg.), 1,600 m² housing for meat production on one unit (Control Bodies standards may vary).

Health See layers.

Finishing To prevent intensive rearing finishing period of at least 81 days (corresponding to traditional free-range definition in the UK) if fast growing strains are used. Slow growing non-organic birds at least 70 days, slow growing organic chicks can be slaughtered at any age. May be hung after slaughter to maximise flavour.

Slaughter Own killing assumed. Low-throughput abattoir, processing 200 birds/ week (10,000 birds/year max.). Hygiene, welfare and marketing regulations must be adhered to and prior discussion with Environmental health is recommend. Costs of £3.00 to £4.50/bird (depending on batch size and regularity) if killed elsewhere, but finding MHS registered slaughter facility willing to do small batches can be difficult.

Other Costs for brooder heating and bedding.

Table poultry (producer-killed and dressed)

Physical assumptions

Finishing period	81 days	10% mortality	2500 birds per ha per year
	3.5 batches per year		8750 birds per ha per year

Financial data

					£/bird	
Table birds	2.0 kg	(4.4 lbs)	@	11.00 £/kg	22.00	
Less one day old chick	1.10 chicks		@	0.95 £/chick	-1.05	
Total output						**20.96**
Feed - starter	1.00 kg/bird		@	810 £/t	0.81	
Feed - finisher	9.50 kg/bird		@	760 £/t	7.22	
Killing and dressing					3.50	
Other livestock expenses					0.75	
Total variable costs						**12.28**
Gross margin excluding forage						**8.68**

Sensitivity analysis

For explanation, see p. 104

	Change in value (+/-)	*Change in GM (£/bird)*	*Value range Low*	*High*	*Gross margin range excl. forage (£/bird)*	
Sale price	0.1 £/kg	0.20	10.00	12.50	6.68	11.68
Finishing weight	0.1 kg/bird	1.10	1.7	2.5	5.38	14.18
Starter concentrate use	0.1 kg/bird	-0.08	0.5	1.50	8.27	9.08
Starter concentrate price	10 £/t	-0.01	750	850	8.64	8.74
Finisher concentrate use	1 kg/bird	-0.76	8.5	10.5	7.92	9.44
Finisher concentrate price	10 £/t	-0.10	650	875	7.58	9.72
Killing & dressing cost	0.1 £/bird	-0.10	3.0	4.5	7.68	9.18

9. Organic Poultry Production for Meat: Technical Guide, Organic Centre Wales (2010): www.organiccentrewales.org.uk/uploads/poultry_guide_english.pdf
OF&G Technical guide TL210 Poultry Slaughter www.organicfarmers.org.uk

10. Blair R (2008) Nutrition and Feeding of Organic Poultry CABI- Publishing Wallingford

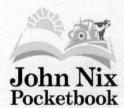

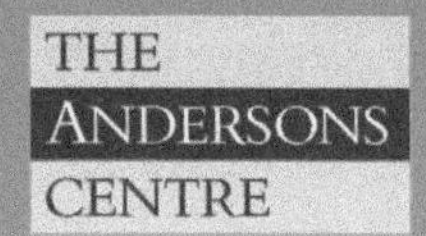

Land

The legislation and markets governing land tenure are not specific to organic farming. Although some attempts have been made to sell land with organic certification at higher prices, these have rarely met with success. Occasionally, private individuals and organisations, e.g., the National Trust and the Ernest Cook Trust, may let land with a specific requirement for organic management. Tenants may be affected by some of the provisions of the Organic Support Schemes (Section 4 & 5) relating to the length of time organic management must be maintained.

The long-term commitment which conversion to organic production entails could also be affected by the Agricultural Tenancies Act of 1995[1]. Farm Business Tenancies established under the Act have greater flexibility with respect to term of tenancy, rent levels and rent reviews, but there is no provision for security of tenure beyond the agreed tenancy period. All tenancies entered into before 1st September 1995 remain subject to the Agricultural Holdings Act 1986. More recent reforms amending the 1985 and 1995 Acts (The Regulatory Reform (Agricultural Tenancies) (England and Wales) Order 2006 (SI 2006 No. 2805)), designed to make it easier for tenant farmers to develop their businesses by restructuring and diversifying, came into force on 19 October 2006[2].

Labour use and costs

Organic farming has typically been considered to have a greater labour use per hectare than conventional. The Organic Research Centre convened a European wide review[3] of literature to identify trends in organic farm labour and farm structure. The review found that presenting results by farm type usually indicate higher labour use per hectare on organic than conventional arable farms. However, similar or lower labour use is reported on organic livestock farms, and the results are mixed for other farm types (see also Section 5).

The main factors contributing to possible increased labour requirements are:

- the increased diversity and complexity of organic systems;

1. www.gov.uk/guidance/agricultural-tenancies
2. See: Guide to the Regulatory Reform Order 2006. Defra, London. www.netlawman.co.uk/ia/agricultural-tenancies-2006
3. Orsini S, Padel S. and Lampkin N, 2018. *Labour use on organic farms: a review of research since 2000.* Organic farming, 4(1), pp.7-15. www.librelloph.com/organicfarming/article/view/of-4.1.7

- additional enterprises within the farm reduce the opportunities for economies of scale and specialisation, but may also permit a better distribution of labour requirements, for example between winter and spring-sown crops;
- the introduction of marketing and/or processing activities to benefit from premium prices and add value to farm products;
- the introduction of labour-intensive, high-value enterprises such as field-scale vegetables and the introduction of livestock on mainly arable holdings.

For many organic arable and livestock enterprises, labour requirements per hectare or per animal are similar to conventional systems. Labour requirements may be reduced on livestock holdings due to lower stocking rates leading to lower total stock numbers. On horticultural holdings the value of the crops produced is closely related to numbers employed and may be a key factor. Additional labour requirements for weed control and harvesting of horticultural crops will depend on the level of mechanisation. Increased labour use for marketing and processing or high-value enterprises is only worthwhile where higher returns justify the cost.

Labour requirements for standard farming practices are illustrated on p. 223 , but other sources should be consulted for more detailed estimates.

Labour costs

FBS survey results for England 2021/22 (see Section 5) show higher costs for paid labour per hectare for organic cropping and mixed farms compared to conventional. Costs per hectare were also a bit higher on LFA grazing farms, but lower for dairy, horticulture and lowland grazing. This does not include any notional values for farmer and spouse labour.

Paid labour costs (as in conventional systems) are underpinned by minimum wage legislation (see below). Actual wages will vary depending on the local labour market and the shortage of labour experience at all levels. Farmers also need to pay national insurance of 13%, employers' liability and should calculate an additional 20% for annual and sick leave.

England: Agricultural and horticultural workers in England must be paid in accordance with the appropriate hourly **National Minimum Wage** rate, but no changes could be made to the terms of existing contractual agreements in place before the abolition of the Agricultural Wages Board (AWB), without mutual consent. In April 2016 the Government introduced the **National Living Wage** for workers over 25. From the 1st of April 2021 the age at which the National Living Wage should be applied was reduced to 23 and over.

National Minimum Wage/National Living Wage rates, from 1st April 2023 £/hour

23 and over	21 to 22	18 to 20	Under 18	Apprentice*
10.42	10.18	7.49	5.28	5.28

* *Aged under 19 or aged 19 or over in first year of apprenticeship*

Wales: The Welsh Government has its own Agricultural Wages Board, setting wages at the same level as the UK National Living Wage. The Agricultural Wages (Wales) Order 2023[4] defines the statutory minimum wages for agricultural workers, both by category and by age.

AWB minimum wage rates for Wales, from 1st April 2023 £/hour

Grade	Description	£/hour
A1	Agricultural Development Worker (16 – 17 Years)	5.28
A2	Agricultural Development Worker (18 – 20 Years)	7.49
A3	Agricultural Development Worker (21 – 22 Years)	10.23
A4	Agricultural Development Worker (23 Years +)	10.47
B1	Agricultural Worker (16 – 17 Years)	5.28
B2	Agricultural Worker (18 – 20 Years)	7.49
B3	Agricultural Worker (21 – 22 Years)	10.23
B4	Agricultural Worker (23 Years +)	10.74
C	Agricultural Advanced Worker	11.07
D	Senior Agricultural Worker	12.14
E	Agricultural Manager	13.32
Apprenticeships		
	Apprentice Year 1	5.28
	Apprentice Year 2 and Beyond (16-17 Years)	5.28
	Apprentice Year 2 and Beyond (18-20 Years)	7.49
	Apprentice Year 2 and Beyond (21-22 Years)	10.18
	Apprentice Year 2 and Beyond (23 Years +)	10.42
Allowances		
	Dog Allowance – per dog per week	9.36
	Night Time Work Allowance – per hour	1.78
	Birth Adoption Allowance – for each child	73.60
	Accommodation Offset Allowance (House)	1.65 per week
	Accommodation Offset Allowance (Other)	5.29 per day

4. https://www.legislation.gov.uk/wsi/2023/260/made

Scotland: The Scottish Agricultural Wages Boards (SAWB) continues to operate, setting wages at the same level as the UK Living Wage. Workers who also hold certain agriculture or horticulture qualifications[5] are also entitled to an additional £1.55/hr. Minimum overtime rates are payable for workers employed for more than 26 weeks with the same employer working for more than 8 hours on any day or for more than 39 hours in any week. Those owning dogs as part of their work are entitled to £8.00/dog each week, up to a maximum of 4 dogs.

The following rates applied from the 1st of April 2023:

SAWB minimum wage rates for Scotland, from 1st April 2023 £/hour

Description	£/hour
All Workers	10.42
SCQF* Level 4 or 5 Apprentices^	6.53

**Scottish Credit and Qualifications Framework or equivalent.*
^Paid for 18 months. Applies to apprentices under 19 years of age and those of 19 years or over in the first year of their apprenticeships.

Northern Ireland: Orders cover payments to qualified and unqualified workers in six main grades. Orders also stipulate overtime rates, minimum rates for apprentices and other allocations such as dog allowances and holiday entitlements.

AWB minimum wage rates for Northern Ireland, from 1st April 2023 £/hour

Grade	Description	£/hour
1	Minimum Rate	7.54
2	Standard Worker	8.13
3	Lead Worker	10.16
4	Craft Grade	10.92
5	Supervisory Grade	11.49
6	Farm Management Grade	12.48

5. SCQF 6/7 or above (including Scottish/National Vocational Qualification at Level 3, National Certificate or Higher National Diploma)

Machinery

Organic farming involves the use of a wide range of machinery, much of which is common to conventional farming. Analysis of fixed costs on organic compared to conventional for different farm types as reported in the Farm Business Survey 2012/22 Organic Farming in England (see Section 5) indicates that machinery costs per hectare are significantly lower for cropping, mixed, dairy and horticulture farms, higher for LFA farms and slightly lower for lowland grazing farms.

Possible savings can be made on sprayers and fertiliser spreaders, but higher costs due to greater enterprise diversity and therefore a larger range of machines being required, notably on mixed farms. However, the variation between farms is considerable. Standard labour and machinery costs for most common farming practices are illustrated on p. 223.

Specialist machinery for weed control such as inter-row cultivators, harrow-combs, spider-tine hoes, brush and flame weeders may be required, particularly if arable crops or vegetables are grown on a larger scale. Purchase prices and operating costs for specialist weed control machinery are shown on p. 224.

Good soil structure is essential for productive organic grassland. The use of sward slitters to aerate the soil and grassland lifters to loosen the soil down to 20 cm will help overcome surface and subsurface poaching and compaction, particularly that caused by machinery or livestock during wet weather.

Machinery Costs

Soil slitter			
Operational width (m)	3	6	
Purchase price (£)	6000	15500	
Soil lifter			
Operational width (m)	2.75		
Purchase price (£)	7500		
Arable crop spring tined weeder			
Operational width (m)	6.4	8.9	
Purchase price (£)	14000	15500	
Grass Harrow			
Operational width (m)	3	9	18
Purchase price (£)	6200	13300	47000
Grass harrow with airseeder			
Operational width (m)	3	6	12
Purchase price (£)	10-11000	13-22000	25-30000
Row crop star hoe (6 row)			
Purchase price (£)	18-20000		
Multi-row rototiller (6 row)			
Purchase price (£)	12-18000		
Camera guided rowcrop hoe			
Operational width (m)	6	12	
Purchase price (£)	45000	65000	
Modular cameleon weeder			
Operational width (m)	6		
Purchase price (£)	90000-100500		
Crimper roller			
Purchase price (£)	10300		
In-row precision weeding hoe			
Operational width	6 (row)	12 (row)	
Purchase price (£)	69500	123000	

Standard labour and machinery costs

Extracted with permission from Nix's 2023 Farm Management Pocketbook (53rd Edition) and NAAC 2022/23 price guide. Contractor prices are a guide of middle prices. The actual price may vary considerably between regions, soil types, distance travelled, size of contract undertaken, size and type of equipment used, amount of product applied etc.

Tractor and labour costs (per hour excluding fuel)	*Farmer*	*Contractor*
Operators	15.35	*incl. labour*
100 - 150 HP, 4wd		47.40
150 - 220 HP, 4wd		50.10
1220 - 300 HP, 4wd		64.60

	Labour hours				*Machinery & labour costs*			
	Average		*Premium*		*Farmer*		*Contractor*	
	per ha	*(acre)*	*per ha*	*(acre)*	*per ha*	*(acre)*	*per ha*	*(acre)*
Ploughing/sub-soiling (incl. fuel)								
Ploughing	1.4	*(0.57)*	1.0	*(0.40)*	71	*(29)*	75	*(30)*
Rotovating, stubble	1.8	*(0.73)*	1.3	*(0.53)*	81	*(33)*	88	*(36)*
Rotovating, grassland	2.7	*(1.09)*	2.0	*(0.81)*	106	*(43)*	110	*(45)*
Subsoiling	1.3	*(0.53)*	0.8	*(0.32)*	62	*(25)*	77	*(31)*
Cultivations								
Stubble cultivating	0.7	*(0.28)*	0.5	*(0.20)*	51	*(21)*	51	*(21)*
Disc harrowing, light	0.7	*(0.28)*	0.5	*(0.20)*	45	*(18)*	57	*(23)*
Disc cultivating heavy	0.8	*(0.32)*	0.5	*(0.22)*	57	*(23)*	61	*(25)*
Spring-tine cultivating	1	*(0.40)*	0.7	*(0.28)*	29	*(12)*	46	*(19)*
Rotavating	3.5	*(1.42)*	2.5	*(1.01)*	81	*(33)*	95	*(38)*
Power harrowing	1.2	*(0.49)*	0.8	*(0.32)*	70	*(28)*	74	*(30)*
Seed-bed harrowing	0.7	*(0.28)*	0.4	*(0.16)*	51	*(21)*	65	*(26)*
Rolling, Cambridge	0.6	*(0.24)*	0.4	*(0.16)*	12	*(5)*	26	*(11)*
Rolling, flat	1.3	*(0.53)*	0.8	*(0.32)*	23	*(9)*	34	*(14)*
Drilling/planting								
Cereal, standard	0.7	*(0.28)*	0.5	*(0.20)*	33	*(13)*	56	*(23)*
Cereal, combi-drilling	1	*(0.40)*	1.4	*(0.56)*	78	*(32)*	81	*(33)*
Roots/Maize	1.5	*(0.61)*	1.5	*(0.61)*			57	*(23)*
Precision drilling, sugar beet	1.3	*(0.53)*	0.8	*(0.32)*	70	*(28)*	61	*(25)*
Broadcast seed	0.6	*(0.24)*	0.4	*(0.16)*	26	*(11)*	32	*(13)*
Potatoes, ridging	1.4	*(0.57)*	1.1	*(0.45)*	70	*(28)*	94	*(38)*
Potatoes, destoning	2.5	*(1.01)*	2.0	*(0.81)*	282	*(114)*	285	*(115)*
Potatoes, planting	3.2	*(1.30)*	2.3	*(0.93)*	182	*(74)*	198	*(80)*
Harvesting								
Cereals, combine/cart	1.1	*(0.45)*	0.8	*(0.32)*	120	*(49)*	134	*(54)*
Beans, combine/cart	1.2	*(0.49)*	1.0	*(0.40)*	97	*(39)*	109	*(44)*
Straw, baling/cart	4.8	*(1.94)*	3.4	*(1.38)*	68	*(28)*	86	*(35)*
Grain, cleaning/drying	0.5	*(0.20)*	0.2	*(0.08)*	57	*(23)*	85	*(34)*
Potatoes, 2 row unmanned	0.1	*(0.04)*	0.1	*(0.05)*	714	*(289)*	695	*(281)*
Topping set-aside/green manure	1.6	*(0.65)*	1.0	*(0.40)*	29	*(12)*	46	*(19)*
Silage, maize (complete service)	1.8	*(0.73)*	1.7	*(0.69)*			190	*(77)*
Grass, mow	1.0	*(0.40)*	0.7	*(0.27)*	37	*(15)*	37.00	*(15)*
Silage, turn	0.7	*(0.28)*	0.5	*(0.20)*	20	*(8)*	21	*(8)*
Silage, forage harvest	2.0	*(0.81)*	1.6	*(0.65)*	79	*(32)*	89	*(36)*
Silage, cart & clamp	5.3	*(2.14)*	4.0	*(1.62)*			148	*(60)*
Hay, turning cost per crop	2.6	*(1.05)*	1.9	*(0.77)*	80	*(32)*	85	*(34)*
Hay, bale (small bales)	1.3	*(0.51)*	0.9	*(0.36)*	92	*(37)*	92	*(37)*
Hay, cart	6.0	*(2.43)*	4.5	*(1.82)*	110	*(45)*	117	*(47)*
Other weed control/crop protection								
Inter-row cultivation	1.6	*(0.6)*	1.0	*(0.4)*	74	*(30)*	77	*(31)*
Harrowing (spring tine)	0.7	*(0.3)*	0.4	*(0.2)*	29	*(12)*	46	*(19)*
Spraying (excl. materials)	0.23	*(0.1)*	0.11	*(0.0)*	10	*(4)*	16	*(6)*

Weed control operational costs

Own estimates & data extracted with permission from Nix's 2023 Farm Management Pocketbook (53rd Edition). Some of the data have been modified or supplemented in the light of other available information.

Spring tine weeder

Operational width (m)	9		12					
Purchase price (£)	13200		14000					
Average speed (12 m machine)			11	km/h =	13.2	ha/h	£/ha	*(£/ac)*
Tine wear	400	ha life	480	tine @	5.35	£/tine	6.4	*(2.6)*
Depreciation	7	yr life	20%	trade in	200	ha/yr	8.0	*(3.2)*
110 HPTractor and	1	operator		@	46.0	£/hr	3.5	*(1.4)*
Total operating cost for	1	pass	with 12 m width				17.9	*(7.2)*

Brush weeders

(2 person steerage weeder)

Operational width (m)	1.8	(35 brush sections)						
Purchase price (£)	14000	-16000						
Average speed	3	km/h	0.54	ha/h			£/ha	*(£/ac)*
Brush replacement	80	ha		@	25	£/brush section	10.9	*(4.4)*
Depreciation	7	yr life	20%	trade in	50	ha/yr	32.0	*(13.0)*
110 HP Tractor and	2	operators		@	46.0	£/hr	85.2	*(34.5)*
Total operating cost for	1	pass					128.1	*(51.9)*

Finger Weeder

Operational width (m)	3	(5 row)						
Purchase price (£)	15000			350 per row for aditional rows				
Operation speed	4	km/h	1.20	ha/hr			£/ha	*(£/ac)*
Finger disk replacement	400	ha	8	fing. @	74	£/disk	1.5	*(0.6)*
Depreciation	7	years	20%	trade in	50	ha/year	34.3	*(13.9)*
1110 HP Tractor and	1	operator		@	46.0	£/hr	38.3	*(15.5)*
Total operating cost for	1	pass					74.1	*(0.0)*

Bed Weeder

Operational width (m)	6							
Purchase price (£)	8000							
Average speed	0.1	km/h	0.06	ha/hr	(depends on weeds)		£/ha	*(£/ac)*
Depreciation	7	year life	20%	trade in	20	ha/year	46	*(19)*
110 HP tractor	1	operator			46.0	£/hr	767	*(310)*
Casual labour		9 people		14 £/hr			2100	*(850)*
Total operating cost for	1	pass					2912	*(1179)*

Flame weeder, infra-red combination (small)

Operational width (m)	2							
Purchase price (£)	24000							
Average speed	4	km/h =	0.80	ha/h			£/ha	*(£/ac)*
Gas	50	kg/h		@	185	p/kg	115.6	*(46.8)*
Depreciation	7	yr life	20%	trade-in	30	ha/yr	91.4	*(37.0)*
75 HP Tractor and	1	driver		@	46.0	£/hr	57.5	*(23.3)*
Total operating cost for	1	pass					264.6	*(107.1)*

Flame weeder, infra-red combination (medium)

Operational width (m)	6							
Purchase price (£)	57000							
Average speed	4	km/h =	2.40	ha/h			£/ha	*(£/ac)*
Gas	120	kg/h		@	185	p/kg	92.5	*(37.4)*
Depreciation	7	yr life	20%	trade-in	90	ha/yr	72.4	*(29.3)*
110 HP Tractor and	1	driver		@	46.0	£/hr	19.2	*(7.8)*
Total operating cost for	1	pass					184.0	*(74.5)*

Combcut weeder

Operational width (m)	6							
Purchase price (£)	16000	(delivered from Sweden)						
Average speed	10	km/h	7	ha/hr	(depends on weeds)		£/ha	*(£/ac)*
Depreciation	7	year life	20%	trade in	80	ha/year	26	*(11)*
75 HP tractor and	1	operator			46.0	£/hr	7	*(3)*
Total operating cost for	1	pass					33	*(13)*

Manure handling and storage

Farm-yard manures (FYM) and slurry represent a valuable resource in organic farming systems that should be planned into the rotation with applications directed at points of greatest nutrient off-take, for example where forage is conserved or sold off the farm. Normal application rates are 10-30t/ha (4-12t/ac) for FYM or 14-20 m³/ha (1,250-1,750 gallons/ac) for slurry. Nutrient budgeting (see p. 90) can be used to check whether returns of manures and/or slurry are adequate to compensate for nutrient removal in a balanced rotation. If mismanaged, farmyard manure represents not only a significant environmental problem, but also a loss of productive nutrients with financial consequences for the farm business.

Equipment and application costs of fertiliser, FYM/Slurry

Own estimates & data extracted with permission from Nix's 2023 Farm Management Pocketbook 53rd Edition. Some of the data have been modified or supplemented in the light of other available information.

Solid manures (FYM/Compost)			*Price range*	
Moving bed rear unloading spreader		10-12 t	18500	27000
		12-14 t	25000	35000
Side delivery (e.g. West)		3-8 t	6000	14000
		8-15 t	14000	30000
Liquid manures (Slurry)				
Slurry tanker spreaders	*(vacuum)*	5-6 cu m	8500	11900
	(low ground pressure)	9-11 cu m	20300	27000
Slurry tanker (7.5m) wide with training shoe		9.5 cu m		35000
Trailing shoe slurry application bar (6m), umbelical system pump and pipe			approx	35000
Tankers can be adapted to spread more evenly using a boom: 7.5m			approx	13700
Tarpaulin cover (excludes rain water), 15 m tank (i.e. 1000 cu m), fitted			approx	31000

Application costs	*Labour hours*				*Machinery & labour costs*			
	Average		*Premium*		*Farmer*		*Contractor*	
	per ha	*(acre)*	*per ha*	*(acre)*	*per ha*	*(acre)*	*per ha*	*(acre)*
Fertiliser	0.4	*(0.16)*	0.3	*(0.12)*	9	*(3.64)*	15	*(6.07)*
Lime spreading incl. lime (5t/ha)							38	*(15.38)*
FYM, load & spread (per hour)					58		60 £/h	
Slurry, tanker (per hour)					60		66 £/h	
Umbilical trailing shoe							135 £/h	
Umbilical top spreading							78 £/h	

Organic standards currently permit the use of brought in non-organic manures, depending on livestock system origin (including feed sources) and treatment, up to the equivalent of 170kg N/ha. Great care has to be taken to avoid pollution

6. Managing Manure on Organic Farms (2002) ADAS and Elm Farm. see orgprints.org/24819;

   IOTA Technical Leaflet (6), A guide to nutrient budgets on organic farms: https://orgprints.org/id/eprint/31654/

by manures or slurries[7]. Storing manure in a field is unsatisfactory due to the risk of nutrient leaching and pollution. This may be minimised by careful siting and covering with a purpose made, impermeable, breathable sheet during winter. A more satisfactory storage system involves three solid side walls with a concrete base and run-off drains into a liquid storage unit (see costings next page). 580mm rainfall over 180 days will result in 72.5 m³ of run-off, requiring either regular spreading or a storage tank of sufficient size.

Composting may be carried out on a similar site, but with two walls, and the subsequent reduction in volume reduces handling costs by half. Making high quality compost is based on controlling the ingredients, regular turning and monitoring of temperature, CO_2 and moisture content (may involve turning 6 –10 times with a specialist turner over an 8–10 week period). Less thoroughly composted material may be used on arable and grassland, turning 2 or 3 times with a fore loader. Costs for composting are in the range of £8-15/t, depending on depreciation costs of the facilities, management of the process and the number of times the material is turned.

Total nutrients in organic and conventional farm yard manures

Description	Production	DM%	N	P_2O_5	K_2O
		(kg/t fresh weight FYM or kg/m³ slurry)			
Cattle FYM (organic)		25	5.9	3.1	6.6
Cattle FYM (conv.)	8.8 t/cow/180 days	25	6.0	3.5	8.0
Cattle slurry (organic)		6.0	2.0	0.8	2.3
Dairy slurry (conv.)	10 m³/cow/180 days	6.0	3.0	1.2	3.5
Beef slurry (conv.)	10 m³/cow/180 days	6.0	2.3	1.2	2.7
Sheep FYM (conv.)		2.5	6.0	2.0	3.0
Poultry–deep litter	53 kg/layer/year	70	10	18	13
Poultry–broiler litter	0.43 kg/broiler/week	70	14.5	22	14
Pig solid (organic)		25	6.5	6.1	6.5
Pigs–solid (conv.)	4.3 t/sow/yr	25	7.0	7.0	5.0
Pigs–slurry (conv.)	0.7-1.4 m³/pig/100 days	4	4	2.0	2.5

7. www.gov.uk/government/publications/protecting-our-water-soil-and-air
www.gov.uk/government/publications/fertiliser-manual-rb209

Manure storage and handling costs

Own estimates & data extracted with permission from Nix's 2023 Farm Management Pocketbook 53rd Edition. Some of the data has been modified or supplemented in the light of other available information.

Manure storage	400 t FYM				Cost (£)
Concrete base 150mm, hard core 150 mm	(25x20)	500 sq m	56.00	£/m²	28000
Concrete block wall 215mm x 1.5m					
incl. strip foundations		25 m @	225.00	£/m	5625
Slatted drains concrete channel		10 m @	8.00	£/m	80
Slats		10 m @	115.00	£/m	1150
Drains, 150 mm		10 m @	38.00	£/m	380
Effluent tank, 20 cu m					12000
Total cost					47235

Composting

Equipment purchase				Cost (£)
Compost turner, tractor driven e.g., CMC ST200		Small scale	2.2 m	23,000
	CMC ST300	Medium scale	3.4 m	40,000
	Fleece roller add			15,000
Compost. Area: Concrete, 3 batches @300t, 50*35 m, drainage to tank			90,000	100,000

Compost operation costs

Creating windrow *rear unload spread*	144 ts (4x2x20m)			
Tractor (90HP) with foreloader	4 hours @	48.00	£/hour	192
Tractor/rear unload spreader	4 hours @	61.00	£/hour	244
Total cost				436
Turning windrow *with foreloader*	144 ts (4x2x20m)			
Tractor (120HP) with foreloader	4 hours @	48.00	£/hour	192
Total cost (per t)				192
Cost of creating windrow and turning 2 times		5.69	£/tonne	820
Compost covers	200 sq m			
Heavy duty 200 micron plastic sheet		1.00	£/m²	200
Toptex compost sheet (4m by 50 m)		4.80	£/m²	960
Compost tea				
Brewing equipment (270 lites capacity)				4,200

In addition, with the Defra targets for reducing land-filled biodegradable waste there continue to be commercial opportunities for farmers to get involved in green-waste composting at a community level. For large-scale operations, a green waste composting site will require planning permission and Environment

8. See also Organic Farmers and Growers technical leaflet TL120 Green Waste Compost. https://assets.ofgorganic.org/tl-120-green-waste-composting.t2ji2q.pdf

Agency approval. For small quantities composted for use on the farm, this may not be necessary (see Section 15 for addresses). Green waste compost does have problems of contamination, particularly plastic.

Slurry storage is in standard tanks or lagoons. A slurry aerator enhances the quality of the slurry by improving the smell and nutrient stability. Bacterial slurry additives may be used in slurry stores with potential reduction in smell and crusting, and improved nutrient availability with claimed improvement in soil life. Cost: £12 for 1 sachet/100 cows/week.

Government targets to meet a reduction in agriculture greenhouse gas emissions will increasingly require the introduction of manure management practices that minimise losses to water and air. The manure storage and handling costs table on p. 227 includes equipment specially to reduce methane and nitrous oxide losses to the air such as cover for slurry lagoon and trailing shoe and injection slurry applicators.

Financial value of farm yard manure

Organic farms may be purchasing manure from other farms, for example where there is a linked farm arrangement with another organic farm or where conventional manure is brought in. The financial value of such manure may be based on the financial value of conventional fertilisers, although there will be other factors including the availability of the nutrients, micronutrients, organic matter and local market demand/supply which will influence the price. The following table provides some guidance, based on well-composted conventional cattle manure from straw bedded yards and using conventional fertiliser costs.

Estimate of financial value of conventional cattle FYM based on average prices for fertilisers

	kg of nutrient per tonne fresh weight	Value of nutrient (£/kg)	Value (£/t)
N	6	£1.88	£11.28
P_2O_5	3.2	£1.41	£4.50
K_2O	8	£1.00	£8.00
SO_3	2.4	£0.43	£1.03
Total			**£24.81**

Reference Nix, J. Farm Management Pocket Book 2023
Note: Fertiliser prices have been particularly high in response to Russia's invasion of Ukraine, prices have begun to normalise since these estimates were produced.

Agricultural waste[9]

Agricultural waste (excluding animal manure and slurry) needs to be recovered or disposed of in ways that protect the environment and human health. Farmers need to either apply for an exemption or take wastes off-farm for disposal at licensed sites. Exemptions are required for:

- **Use of wastes**, such as spreading mushroom compost on your land to improve the soil or hardcore for tracks.
- **Treatment of waste**, such as using an anaerobic digester to help you manage manures and slurries.
- **Disposal of waste**, such as burning hedge trimmings in the open or spreading dredging on the banks of farm ditches.
- **Storage of waste**, such as storing sewage sludge before spreading it under the Sludge (Use in Agriculture) Regulations.

New regulations for waste exemptions will be in force from 2024 and further information from the Environment Agency (England)[10], Natural Resources (Wales), Scottish Environment Protection Agency or DAERA (Northern Ireland)[11].

The National Fallen Stock Company (NFSCo) is a not for profit, farmer led organisation dedicated to delivering a national service for the collection and disposal of fallen stock that farmers use from choice. Membership provides disposal of fallen livestock which can no longer be burnt or buried on-farm. Members will receive a monthly statement from NFSCo summarising the categories and quantities of fallen stock collected. Monthly admin fee of £1.95 for members (£1.45 for those opting for email invoicing) applies for month in which the service is used. Details are available at: www.nfsco.co.uk

Buildings and other capital assets

- In organic farming, there may be additional costs relating to investments in farm buildings and other facilities, for example:
- increased housing space per animal, or conversion of housing from individual penning, or slatted systems to straw yards if required by organic production standards;

9. See www.netregs.org.uk/legislation.aspx for an overview in Scotland and Northern Ireland.
10. www.environment-agency.gov.uk/business/sectors/32779.aspx
11. www.daera-ni.gov.uk/topics/waste

- fencing of fields and supplying water to enable best management of grassland for grazing animals;
- improvements in manure/slurry storage and handling (see previous section);
- facilities for crop storage (e.g., cold-stores), processing and on-farm retailing (if justified by potential returns);
- establishment of new livestock enterprises on arable farms, including investments in fencing, water supplies, housing and, livestock as well as possible extra staff housing.

At the same time, there may also be opportunities for better utilisation of existing equipment and buildings by ensuring that enterprises are complementary in their requirements. The need for produce storage and handling facilities may also be reduced due to lower levels of production.

Some banks (in particular Triodos) actively support the organic sector and provide loans that are specifically targeted at organisations or businesses that aim to generate environmental, social or cultural value. Areas considered suitable for such loans include various enterprises (e.g., arable, dairy, meat, poultry, forestry, horticulture) and investments in marketing (e.g., shops, butcheries, food processing) as well as renewable energy (e.g., wind, solar, hydro, biomass), other eco-development (e.g., shared workplaces, property development) nature development and environmental technologies (e.g. recycling, transport).

There is little evidence from organic farm survey data (see Section 5) that property and capital costs differ significantly between organic and conventional farms. Costs for standard design farm buildings are shown here, adapted to take into account the minimum space requirements of organic production standards. Further modifications, for example environmental enrichment, are desirable on animal welfare grounds (see Section 9), but the costs are not included here.

Capital grant schemes

Some capital grant aid may be available as part of Rural Development Programmes for enhancing farm businesses and supporting existing policies for farm diversification and to encourage collaboration and innovation across the agri-food chain, the introduction of new enterprises, alternative crops and livestock and processing of non-food farm products. Provisions vary between the different parts of the UK.

Building costs

Extracted with permission from Nix's 2023 Farm Management Pocketbook 53rd Edition.
Some of the data have been modified or supplemented in the light of other available information.

Complete Buildings

Fully Covered and Enclosed Barn

Portal frame, 18m span, 6m bays, 6m to eaves, 3m high blockwork walls with sheet cladding above, 150 mm thick concrete floor (not incl. doors & indoor fittings)	270 £/m²

Cows and Cattle Housing

1. Covered strawed yard, enclosed with ventilated cladding, concrete floor, pens only. Organic standards require minimum bedded area 1m²/100kg liveweight plus 0.75m² inside or outside exercise area	975-247 £/head
2. Extra to 1 for 4 m wide double-sided feeding passage, barrier and troughs	400 £/head
3. Kennel building with feed passage	1400 £/head
4. Portal framed building with cubicles	1950 £/head
5. Extra to 4. for feed stance, feeding passage, barriers and troughs	800 £/head
6. Loose box 16 m² floor area laid to falls, rendered walls	450 £/m²

Sheep Housing

1. Penning, troughs, feed barriers and drinkers installed in suitable existing building	45 £/ewe
2. Purpose-built sheep shed Timberframed, with concrete block walls to 1.8, boards above, internal feed passages, all service. Organic standards require minimum 1.5m²/ ewe	300 £/ewe

Pig Housing

1. Yards with floor feeding (7.5m²/sow)	485 £/sow
2. Dry sows arcs	750 £ each
3. Farrowing sow arcs	650-800 £ each
4. Outdoor equipment (ex weaners): Arcs, fencing, troughs etc.	600-1000 £/sow

Poultry Housing

1. Mobile house (2-300 birds)	67 £/bird
2. Free range: stocking rate of 620/ha (250/acre) and 500/house, housing designed for maximum 6 birds per m²	54-66 £/bird

(Cost varies with size of unit and degree of automation)

Storage

Silage	*£/t stored*
1. Timber panel clamp on concrete base with effluent tank	145
2. Precast concrete panel clamp with effluent tank	325

Slurry			*£/m³ stored*
1. Lined lagoon with safety fence			60
2. Glass-lined steel slurry silo	small	400 m³	50
	medium	1200 m³	72
	large	3600 m³	65

TRINITY NATURAL
CAPITAL PRO
COUNCIL

LAUNCHED 7 SEPTEMBER 2023, WINDSOR, UK

VISIT WEBSITE

SECTION 13: ENVIRONMENTAL MANAGEMENT

SECTION SPONSOR: TRINITY NATURAL PRO COUNCIL WITH TRINITY NATURAL CAPITAL GROUP

Environmental management and organic farming

Positive environmental management is a key aspect of organic farming. Production standards (Section 3) incorporate a number of environmental protection requirements and many other restrictions have indirect, positive environmental impacts[1]. Habitat management plays an important role in crop protection (Section 5) as well as contributing to biodiversity. Environmental factors need to be considered in detail as part of conversion planning (Section 4) and on-going organic farm management[2].

Organic farming aims to create an integrated, agri-environmental production system, which is self-regulating in terms of pest and disease control and conserves wildlife and landscape. This is best achieved through initial knowledge of the natural habitats occurring on the farm and developing an understanding of how these features, as well as newly created ones and the cropped areas themselves, can be appropriately managed within the production system to conserve species and establish a stable long-term agricultural system. This approach can be achieved through practice and is recognised by a significant number of organic farms winning environmental management awards around the country.

Whole Farm Conservation Plans, including annotated maps, habitat management and creation guidelines, are valuable and practical tools to help farmers achieve these aims. If planning a new conversion (Section 4), conservation and conversion plans can be commissioned through organic advisers working in conjunction with environmental consultants. Farmers in Wales may be able to use services such as training and one to one on-farm sessions, offered under Farming Connect[3] for integrating conservation issues with strategic business planning. Practising organic farmers may already have established local sources of conservation advice through various environmental agencies (see below).

1. Lampkin N, Pearce B (2021) *Organic Farming and Biodiversity: Policy Options*. IFOAM Organics Europe, Brussels. https://read.organicseurope.bio/publication/organic-farming-and-biodiversity/

   Lampkin N et al. (2015) *The Role of Agroecology in Sustainable Intensification*. Report for Scottish Natural Heritage. Organic Research Centre, Newbury. https://orgprints.org/id/eprint/33067/
2. *Wildlife and biodiversity: integration and management of farming and wildlife for their mutual benefit*, IOTA Research Review, https://tinyurl.com/IOTA-biodiversity
3. https://businesswales.gov.wales/farmingconnect/

If so, the aim should be to integrate advice and information to develop an overall organic farming and environmental management plan, to benefit all aspects of the system and the environment.

The conservation adviser needs to be aware of the inter-relationship between conservation and organic farming and understand that conservation is applicable to the whole farm, including the cropped areas. Conversely, the management of the non-cropped areas, for example as predator refuges, is of key importance to the functioning of the organic system. Therefore, the main objectives of a conservation plan in organic farming could be summarised as the identification and use of appropriate, site-specific, organic management practices to:

- maximise the wildlife and landscape value of the farmed and non-farmed areas of the farm;
- maximise the benefits to the farmland production systems from non-cropped areas;
- maintain and enhance existing wildlife habitats, biodiversity and landscape features;
- create new habitats to provide wildlife value and enhance amenity and farming aspects, e.g., new ponds, hedges, shelter belts and beetle banks, using indigenous local varieties and species as far as possible.

Environmental advice and further information

Advice on conservation management and grant eligibility is available, in some cases free of charge, from:

- Farming and Wildlife Advisory Group (FWAG) www.fwag.org.uk/
- DAERA Northern Ireland Helpline: 0300 200 7842
- Game and Wildlife Conservation Trust www.gwct.org.uk/farming/advice
- Natural England www.gov.uk/government/organisations/natural-england
- Natural Resources Wales https://tinyurl.com/NatRes-Wales
- Nature Scot www.nature.scot/
- Teagasc Republic of Ireland www.teagasc.ie/environment
- SAC Consulting and other consultancy organisations (see Section 15) www.sruc.ac.uk/business-services/sac-consulting/

Contact details for EIA applications

England (Natural England) https://tinyurl.com/NaturalEngland-EIA

Scotland (NatureScot) https://tinyurl.com/NatureScot-EIA

Wales (Welsh Government) https://tinyurl.com/Wales-EIA

Northern Ireland (Department of Agriculture, Environment and Rural Affairs) https://tinyurl.com/NIreland-EIA

Nitrate Vulnerable Zones (NVZ)

NVZs are areas of land at particular risk from agricultural nitrate pollution. Zones are reviewed and updated every four years to account for changes in water pollution. The latest NVZs in England were established in January 2020 and will remain until 2024. Around 55% of land in England is within an NVZ. NVZ measures for Wales began in April 2021; further compliance requirements came into force 1st January 2023 and more are forecasted for August 2024[4].

Maps outlining current NVZs in England, Scotland and Wales are published on respective government websites. Defra sends written notification to owners and occupiers of land within an NVZ. Farms within these areas must comply with Action Programme measures, including limiting and controlling nitrogen fertiliser use to crop requirement only, limiting and controlling the spreading of manures, storing slurry (where this is required to comply with closed periods for spreading certain types of manure); and keeping farm records. It is likely that producers following the standards for organic production can comply with these management restrictions as they are largely based on good agricultural practice.

By the end of April each year, farms within NVZs must be able to show records relating to slurry production and storage for the five-month storage period for cattle slurry. These have to include up-to-date import/export details, along with sites and duration of use of farmyard manure heaps. Records must also include the whole farm limit for the amount of N produced during the previous year and for the number and type of livestock kept on farm. Farmers can apply for a grassland derogation whereby eligible farms could use up to 250kg of N per hectare from grazing livestock manure. The grassland derogation application window opened in June 2023 but the window for application could vary each year, so it is advised to contact the Environment Agency before applying.

4. https://tinyurl.com/Wales-water-guidance

Agri-environment/Environmental Land Management grant schemes

Agri-environment and Environmental Land Management (ELM) schemes are available in each of the UK administrations and in the Republic of Ireland.

The current grants supporting conversion to and maintaining organic farming (Sections 4 and 5) are implemented as part of existing Countryside Stewardship in England, AECS in Scotland, the Organic Conversion Scheme and Glastir in Wales, and the Environmental Farming Scheme in Northern Ireland.

As with all support schemes linked to retained and/or updated EU legislation, their long-term future remains unclear following the UK's exit from the EU. The roll out and development of post-EU exit agri-environmental funding measures will likely continue in each of the UK's administrations until 2025. However, there are a number of current agri-environmental payments that are cross-compatible with current support for organic farmers' conversion and maintenance.

England: Countryside Stewardship

The Countryside Stewardship (CS) scheme provides financial incentives for farmers and land managers to adopt environmental measures such as: habitat conservation and restoration, flood risk management, woodland creation and restoration, water pollution reduction, and hedgerow and boundary management. The scheme is competitive, and applications are scored against local priorities. There are also a number of grant options for uplands including management of rough grazing, management of grazing for birds and management of moorland.

Funding through the CS scheme is available at three grades: **Mid-Tier, Higher Tier** and **Capital Grants**. Most Mid and Higher Tier agreements run for up to 3-5 years with agreements starting on 1st January each year and new applications opening from 5th January in the year of publication (2023). **Higher Tier** grants cover the environmentally significant sites, commons and woodlands and due to their more complex nature, involve working alongside an advisor from Natural England or the Forestry Commission. As of 2023, 20 Higher Tier capital grant options are available (8 actual cost items, and 12 fixed cost items). **Mid-Tier** options are aimed at wider environmental improvements with over 157 options available in the 2023 scheme, including for organic conversion and maintenance (see Sections 4 and 5 respectively). Most, but not all, of the options can be combined with organic conversion and maintenance options – it is advised that this is checked with the Rural Payments Agency before applying.

Unlike the organic conversion and maintenance options, which are not subject to geographical targeting, most CS options are normally subject to some form of environmental targeting, which means they are not available to producers in low-priority areas.

Five specific organic options aimed at enhacing the environmental impact are available for rotational land in both Mid and Higher Tier agreements, which can only be used on land certified as organic or in-conversion:

- *OP1 Overwintered stubble*: £176/ha on stubbles following cereals (not maize), linseed or oilseed rape in order to provide winter feed for birds. Stubbles must be maintained (not topped, grazed or cultivated) until 15th February of the following year, with a green over-wintered cover crop established on at least 10% and not more than 50% of the area. Soils at risk of erosion are not eligible.
- *OP2 Wild bird seed mixture*: £768/ha for establishing a balanced seed mix of at least 6 small seed-bearing plants from a specified list. Blocks must meet minimum size criteria and be re-established every year for annual mixes and every other year for two-year mixes to maintain seed production.
- *OP3 Supplementary feeding for farmland birds*: £887/t for every 2 hectares of wild bird seed mixture, but only if OP2 also included in the agreement. Prescriptions cover timing, frequency and number/location of feed stations, and specify a minimum mix of 70% cereals and 30% small seeds by weight, with no small seed mix exceeding 50% of the total small seed component by weight.
- *OP4 Multi species ley*: £115/ha for establishing multi-species ley including at least 5 grasses (max 75% of seed mix by weight), 3 legume and 3 herb species on arable land or grassland sown to grass for less than 7 years. Management prescriptions apply.
- *OP5 Undersown cereal*: £306/ha for establishing an autumn or spring sown cereal crop (not maize) undersown by 30th April with a grass/flower-rich legume ley.

Capital grant agreements allow applicants three years to complete capital works and for 2023 the value of grants awarded is not capped. Capital items that can be applied for fall under the following four groups: boundaries, trees and orchards; water quality; air quality; and natural flood management. Capital grants are available for producers to apply for all year round and can complement other Mid-Tier and Higher Tier arrangements so long as the same work is not being paid for twice.

Further information on the scheme is available on the website below and information on individual options, supplements and capital items can be found using the CS grant finder tool[5]. The CS schemes are run jointly by Natural England, Forestry Commission England and the Rural Payments Agency. Initial enquiries should be directed to Natural England:

- Email: enquiries@naturalengland.org.uk
- Tel: 0300 060 3900
- https://tinyurl.com/CS-payments-ELM

England: Sustainable Farming Incentive (SFI)

The SFI payment scheme has been introduced post-UK exit from the EU to provide for universal measures needed across the farming sector and is intended to be a main pillar of future farming in the transition away from BPS. As with updates to existing CS support, the full range of offers for SFI are expected to be available by the end of 2024 with the latest round of options made available in June 2023 . Information on which organic CS schemes will be compatible with claiming SFI can be found in summaries provided by in the following websites:

- Government guidance on claiming for SFI and other funding: https://tinyurl.com/SFI-agreements
- Organic Farmers and Growers (OF&G) Summary for organic farmers: https://tinyurl.com/OFG-SFI

The new SFI program has been designed with the intention of offering farmers and land managers greater flexibility to pick and choose the right options for their land. Under the new SFI scheme, the amount of land entered under a single scheme and the number of schemes joined is up to the discretion of farmers and land managers. Should farmers wish to increase the amount of land under a single option or the number of options they apply for, this can be done annually. There is assurance that farmers and land managers will be able to claim SFI payments alongside CS support for organic conversion/maintenance. However, it will likely not be possible to claim for SFI payments on land which farmers and land managers have under another CS option, and it is advised that farmers check the compatibility of an SFI scheme with the existing CS schemes before applying. For the full range of SFI and CS options that are available for organic farmers, see: https://tinyurl.com/DEfra-SFI-CS-Organics

5. https://www.gov.uk/countryside-stewardship-grants

Within SFI there are a number of options that complement existing CS organic conversion and maintenance, and in some cases could make up for the loss of Basic Payment Scheme funding[6]. See also Section 5.

SFI Options available in addition to organic conversion/maintenance

Code	Action	SFI
SAM3	Herbal leys	£382/ha
SAM2	Legumes on improved grassland	£102/ha
NUM3	Legume fallow	£593/ha
IPM3	Companion crop on arable and horticultural land	£55/ha
IPM4	No use of insecticide on arable crops and permanent crops	£45/ha
SAM2	Multi-species winter cover crops	£129/ha

England: Catchment Sensitive Farming (CSF)

The CSF scheme is run by Natural England alongside the Environment Agency and Defra. It seeks to work with farmers in ways that protect water, soil and air. Its assistance comprises of locally informed on-farm advice. Advice is available for the following themes: soil management; nutrient, slurry, and manure management; ammonia emission reduction; farm infrastructure and machinery set-up; pesticide handling; water resources and natural flood management; local environmental priorities; land management; and agricultural transition, including grants.

Details of the CSF programme and a directory for regional CSF programme advisors can be found here: https://tinyurl.com/NE-CSF-advice

Scotland: Agri-Environment-Climate Scheme

The Scottish RDP offers a variety of schemes aimed at meeting the environmental priorities of protecting and improving the natural environment and addressing impacts of climate change. The **Agri-Environment Climate Scheme** (AECS) offers competitive funding opportunities for a range of land management practices that protect natural heritage, improve water quality, manage flood risk and mitigate and adapt to climate change, as well as targeted and non-targeted support for capital items. Under the scheme there are options for both organic conversion and maintenance (see Sections 4 and 5). The scheme supports environmental management options for arable, grassland, upland, peatland, moorland and heath, wetland and bog, farm habitats, small units, control of

6. https://ofgorganic.org/docs/sfi-and-cs-for-organic-farmers-sept-2023.pdf

invasive species, water quality and flood risk. Schemes are regionally targeted, so options available on a portion of land may differ each year. Most applications will require a Farm Environment Assessment (for which there are also funding options available) and agreements are for 5 years. The scheme is delivered by the Rural Payments and Inspections Division (RPID) and Scottish Natural Heritage (SNH) and is expected to be open for applications between January and April each year. Support and advice are available through RPID local area offices[7] or SNH[8] and following submission each project is assigned a case officer.

Organic Farmers are able to claim for a small number of other AECS schemes depending on the sector. However, double funding on similar schemes is not permitted. The Scottish Government provides a downloadable compatibility checker which can be accessed online at this address:
https://tinyurl.com/SG-agri-env-doublepay

For an overview of Scottish grant schemes, the following address has a full list and is regularly updated: www.ruralpayments.org/publicsite/futures/topics/all-schemes

Wales: Glastir sustainable land management schemes

Glastir has in recent times been the most relevant to organic farmers, but ends in 2023. The scheme offered a range of financial support for activities aimed at addressing climate change, improving water management and maintaining and enhancing biodiversity. Glastir Organic (part of Glastir) was a 5-year programme that offered specific options for organic conversion and maintenance (see Sections 4 and 5). The Glastir scheme also included a range of other land management options, normally combinable with Glastir Organic, aimed at the delivery of specific environmental goods and services. The Glastir scheme is scheduled to close in December 2023. This will be replaced by an Interim Agri-environment scheme, running through 2024 focused on habitat management. which will sustain schemes relevant to organic farmers, until the anticipated Sustainable Farming Scheme is implemented in 2025. The scheme will be open to all eligible farmers, including former Glastir Advanced, Commons and Organic agreement holders.

Smaller Glastir schemes, such as the Small Grants Scheme and the Woodland Creation Scheme, have been available to farmers and land managers throughout 2023. At the time of publication most Glastir schemes were closed

7. https://tinyurl.com/SG-RPID-book
8. www.snh.gov.uk/land-and-sea/srdp/srdp-contacts/

for application for the year, and no new information regarding funding in 2024 was available. Until a full document is released outlining the overall plan for environmental funding in 2024, Welsh farmers and land managers can view the most up-to-date application windows on the Glastir pages of the Welsh Government website: www.gov.wales/rural-schemes-application-dates

Northern Ireland: Environmental Farming Scheme

The Environmental Farming Scheme (EFS) offers participants a 5-year agreement to deliver a range of environmental measures and has three levels:

- *Higher Level*, primarily for environmentally designated sites and other priority habitats. EFS Higher is now closed for applications for 2023, but it is expected to reopen in 2024, although at the time of writing no specific fixed dates for this were available.
- *Wider Level* to deliver benefits across the countryside, outside of environmentally designated areas. No longer available for new applicants.
- *Group Level* to support co-operative action by farmers in specific areas such as a river catchment.

The EFS is a single funding scheme, meaning that those currently on one EFS will not be able to apply for a second. As with other UK administrations, 2024 will be a transition year in Northern Ireland whilst a new funding structure for Northern Irish agriculture is developed and implemented. Consequently, EFS wider schemes are not running as well as a number of other environmental schemes such as the Small Woodland Grant Scheme or the Farm Woodland Scheme. To track available agri-environmental grant funding and key dates for application, see: www.daera-ni.gov.uk/grants-and-funding.

Republic of Ireland: Agri-Climate Rural Environment Scheme (ACRES) and other relevant CAP schemes.

Organic farmers can apply for funding for organic conversion and management through the newly introduced ACRES Organic Farming Scheme – of which the first tranche of funding was opened in January 2023. The details and payment rates of the ACRES Organic Farming Scheme are provided in Sections 4 and 5 of this Handbook.

Organic farmers in the Republic of Ireland will also be eligible to apply for other funding opportunities available through ACRES either individually through ACRES General, or collectively with other farms via ACRES Co-operation. There are a number of other agri-environmental payments that could be of relevance to farmers already partaking in the ACRES Organic Farming Scheme. These include:

- The Eco-Scheme[9] will run as part of the CAP Strategic Plan (2023-2027) and provides funding for a range of targeted on-farm practices that benefit the

environment, climate, water quality and biodiversity. Funding windows occur annually with the 2023 window closing on 29th May.

- Sector-specific schemes also provide further agri-environmental funding opportunities for organic farmers in the Republic of Ireland. These include schemes like the Protein Aid and Protein/Cereal Mix (50/50) Crop Scheme[10], or the Suckler Carbon Efficiency Programme[11], which are also compatible with the Organic Farming Scheme. For a comprehensive list of opportunities under the new CAP programme, visit: https://tinyurl.com/IOA-newCAP-organic.

Trees and hedges on farms

Trees and hedges provide a number of indirect economic and environmental benefits including shelter for livestock, windbreaks, stockproof boundaries, habitat for wildlife and beneficial insects, recycling of nutrients from deeper parts of the soil profile via leaf litter and protection against soil erosion. In some situations, direct economic benefits may also be derived from timber, short-rotation coppicing for bioenergy, and fruit/nut production. The Soil Association operates a range of accredited forestry certification schemes to enable farmers to demonstrate environmentally responsible forestry.

The main costs associated with hedgerows are establishment and maintenance costs. Financial support for hedgerow and woodland establishment and maintenance is available under a range of schemes in different parts of the UK:

- *England*: Countryside Stewardship Mid-Tier, Higher Tier, and small capital grant options[12] and Sustainable Farming Incentive hedgerow options
- *Northern Ireland*: DAERA Forestry Grant scheme[13]
- *Republic of Ireland*: Agri-Climate Rural Environment Scheme – Planting New Hedgerows and Planting Trees in Riparian Buffer Zones[14]
- *Scotland*: Agri-Environment Climate Scheme – Creation of Hedgerows[15]
- *Wales*: Glastir Woodland and Glastir Small Grant schemes[16].

9. https://www.gov.ie/en/service/e5ed0-eco-scheme/
10. https://www.gov.ie/en/service/a6431-protein-aid-and-proteincereal-mix-crop-scheme/
11. https://www.gov.ie/en/service/413bc-suckler-carbon-efficiency-programme-scep/
12. For a summary of relevant CS grants see:https://hedgelink.org.uk/grant/countryside-stewardship/
13. www.daera-ni.gov.uk/articles/daera-forestry-grants
14. https://tinyurl.com/ie-ACRES-spec
15. https://tinyurl.com/SG-agri-env-hedge
16. www.gov.wales/small-grants-environment-hedgerow-creation-general-rules-booklet-html

Establishment and maintenance costs of hedges and woodland

Description	Detail	Cost (£)
Hedge planting		
Transplants		0.50-2.00 each
Spirals and canes		0.20-1.00 each
Mesh guards		0.50-2.00 each
Tubes plus stake		1.90 each
Canes		0.30 each
Labour to plant and guard		3.50-5.00/m
Fencing incl. Labour	stock proof	6.00-7.00/m
Fencing incl. Labour	stock and rabbit proof	10.00-12.00/m
Hedge coppicing		
By hand, 2 men and chain saw	100m/day	7.50-9.50/m
Contractor, tractor mounted saw	12.5m./hour (incl. driver, 2 men)	4.60/m
Hedge laying (depends on thickness)	(every 8-20 years)	
Contract labour (excluding stakes)	20-40m/day	17.00/m
Amenity tree planting	half acre block or less	
Transplants		1.80 each
Shelter plus stake and tie average	60cm tree shelters & stake	1.80 each
Trees/day/man	farmer 200, contractor 400	
Shelter belts		
Large species (oak, lime etc.)	100 per 100 m at	0.65 each
Medium species (cherry, birch, etc.)	66 per 100 m at	0.90 each
Shrubs	100 per 100m	0.45 each
Windbreaks		36.00/ha
Woodland establishment (incl. trees)	Contractor	£/ha
Conifers at 2m (i.e. 2,500 tree/ha)	Without guard or fence	1800
Broadleaves at 3m (1,100 ha)	Without guard or fence	1200

Source: The John Nix Farm Management Pocketbook 2023

In addition to government-funded grants, the **Woodland Trust** provides a range of financial and information support for tree planting on farms, including:

- *MOREwoods*: Advice and financial support to create an area of new woodland over 0.5 hectare.
- *MOREhedges*: scheme can cover up to 75% of the cost if you plant 100 metres or more of new hedging and allow a large tree to grow every six metres.

- Further regional support options are available in England, Northern Ireland, Scotland and Wales (Coed Cymru).

For further information on Woodland Trust support and advice, contact:

- Helpline: 0330 333 5303
- Email: plant@woodlandtrust.org.uk
- https://tinyurl.com/WoodlandTrust-grants

Agroforestry[17]

Agroforestry integrates trees and shrubs with crops and/or livestock production, building on the idea of ecological design to optimise beneficial interactions between the woody and other components. These interactions can lead to higher productivity compared to conventional monoculture systems and provide a wide range of services including soil management, microclimate modification, weed control, natural fencing, carbon sequestration and nutrient recycling as well as shelter and feed for livestock. It is particularly relevant for organic producers as agroforestry systems reduce the need for agrochemical inputs by minimising nutrient losses and maximising internal cycling of nutrients, and by enhancing biological control of pests and diseases.

The inclusion of trees at a low density in agricultural land challenges the conventional specialisation of forestry and agriculture in policy mechanisms. This lack of policy support has been one of the main barriers to wider adoption of agroforestry. Yet recognition of the role of trees in meeting the current climate change and nature crises is prominent in a number of key policy documents: The Land use: Policies for a Net Zero UK (2020), Sixth Carbon Budget Policy document (2020), Net Zero Strategy (2021), Environment Act (2021) and England Trees Action Plan (2021–2024) all have tree planting targets. The government has committed to increasing tree-planting rates across the UK to 30,000 hectares per year by March 2025, equating to 90-120 million trees each year (depending on planting density). The Nature for Climate Fund Tree Programme has £753 million to fund tree planting through new woodland creation partnerships with local authorities and charities and to provide landowners with grants and advice to increase woodland creation, expansion and management.

In the current agricultural transition following the UK exit from the EU, the Basic Payment Scheme (the equivalent of the EU Pillar 1 payments) will be phased out

17. Smith, J. (2010). Agroforestry: Reconciling Production with Protection of the Environment: A Synopsis of Research literature. https://orgprints.org/id/eprint/18172/

by 2027. There is exciting potential for agroforestry to be financially supported through the replacement schemes that are in development in the devolved nations. At the time of preparing this current edition of the Handbook, these schemes are still in consultation and final development and so the exact nature of the support cannot be included here.

In **England** BPS is being replaced by the Environmental Land Management (ELM) scheme, with three tiers that offer potential for agroforestry support. Tier 1 (the Sustainable Farming Incentive) – will pay for management changes directed at improving environmental performance of farming operations. Tier 2 (Countryside Stewardship) will pay for management of land specifically for environmental purposes. Tier 3 (Landscape Recovery) will pay for large-scale environmental change including restoring wilder landscapes and large-scale tree planting.

Agroforestry actions under the SFI for establishing and maintaining silvo-arable and silvo-pasture is under development and is to be piloted in 2024. Actions for hedgerows and wood pastures (as well as farm woodlands) are already available and complement existing actions under Countryside Stewardship.

CS and SFI Opportunities for Agroforestry

Code	Action	Payment rate	SFI/ CS
BE3	Management of hedgerows by rotational cutting and leaving some hedgerows uncut	£10 per 100m for 1 side of a hedge	CS
HRW1	Assess and record hedgerow condition	£3 per 100m – one side	SFI
HRW2	Manage hedgerows	£10 per 100m – one side	SFI
HRW3	Maintain or establish hedgerow trees	£10 per 100m – both sides	SFI
BE4	Management of traditional orchards	£264/ha	CS
BE5	Creating traditional orchards	£373/ha	CS
BE6	Veteran tree surgery	£379 per tree	CS
BE7	Supplement for restorative pruning of fruit trees	£113 per tree	CS
WD6	Creation of wood pasture	£544/ha	CS
WD10	Management of upland wood pasture and parkland	£212/ha	CS
WD11	Restoration of upland wood pasture and parkland	£371/ha	CS

HRW1 is available for all hedges even if in a CS scheme. HRW2 (manage hedgerows) cannot be claimed on hedges in Stewardship BE3 but works with all other SFI options and requires a lenient cutting system such as 3-year cutting regime or 2-year if cutting in January or February. HRW3 aims to ensure one tree on average per 100m, which could be a new tree, a sapling or an existing tree.

The England Woodland Creation Offer (EWCO) provides relatively generous funding for woodland scale planting, with opportunities for supporting small group planting, shelter belts and riparian strips. However, the scheme does not support in-field tree planting and no agricultural activity is allowed in EWCO funded woodland, which collectively rules out many agroforestry systems.

In **Wales**, the Sustainable Farming Scheme is in development. This carries the intention to support the target to create 43,000 hectares of new woodland by 2030 to help mitigate climate change, including through agroforestry and hedge planting. Scheme entrants will be required to carry out Universal Actions, which include having at least 10% tree cover on plantable area of their farm, managed in line with the UK Forestry Standard. Farmers wishing to plant more trees than the minimum 10% coverage can receive further support to plant more individual trees, hedges, groups of trees, shelter belts and riparian strips.

A three-year funding package was announced to support the transition to the Sustainable Farming Scheme. As part of this, the Small Grants – Woodland Creation Scheme is aimed at farmers and other land managers to encourage planting of small areas of trees on land which is agriculturally improved or low environmental value in Wales[17]. Funding is available for tree planting to create shelterwoods, alongside watercourses, and in field corners or small fields for stock shelter, biodiversity and woodfuel. The scheme offers 12 years of payments for Maintenance and Premium payment in respect of the new planting. Funding is available for planting areas between 0.1 and 1.99 hectares.

In **Scotland**, the Forestry Grant Scheme includes an agroforestry option, providing support to help create small scale woodlands within sheep grazing pasture land (silvo-pastoral system) or on arable land (silvo-arable system), for the purposes of providing shelter for livestock, providing timber, increasing biodiversity, enhancing the landscape and contributing to Ecological Focus Areas (in specific situations). Payment of an initial grant for planting of trees and an annual payment for five years at two rates, for 400 or 200 trees per hectare, on land used for sheep grazing or arable land. Other grants are available for woodland improvement and sustainable management of woodlands.

18. https://tinyurl.com/Wales-smallgrants-wood

Scottish Forestry are introducing from summer 2023 four new measures to attract more agroforestry planting under the Forestry Grant Scheme. These are:

- Increasing the grant rate for agroforestry projects by 50% from £3,600 per hectare to £5,400 per hectare;
- Making agroforestry funding available for planting fruit and nut and native trees;
- Allowing additional protection measures for trees, to allow cattle to graze within agroforestry projects; and
- Giving farmers more opportunity to participate in agroforestry by adapting the planting thresholds.

In **Northern Ireland**, the Environmental Farming Scheme (EFS) offers participants a 5-year agreement to deliver a range of environmental measures at three levels: a Higher Level, primarily for environmentally designated sites and other priority habitats; a Wider Level to deliver benefits across the countryside, outside of environmentally designated areas; and a Group Level to support co-operative action by farmers in specific areas such as a river catchment. The measures supporting agroforestry are shown below.

EFS Opportunities for Agroforestry

Measure	Capital payment	Annual payment
Establishment of agroforestry systems	✓	Yrs 2-5
Creation of a traditional orchard	✓	Yrs 2-5
Planting of new hedgerows including two protective fences	✓	
Hedge laying, including two protective fences	✓	
Creation of riparian buffer zones with native trees. Options for 2 m-wide zone or 10 m-wide zone available.	✓	Yrs 2-5
Planting native tree corridors (shelter belt or downwind of farm buildings to capture gaseous emissions)	✓	Yrs 2-5
Natural regeneration of native woodland (a grant for an area where grazing is excluded, and where the carbon sequestration and biodiversity value of the site will be enhanced).		Annual grant for 5 yrs

Agroforestry impact on production

Careful design can avoid completely or reduce impacts on crop production, however in-field trees can have a negative impact due to water, nutrient and especially light competition between the trees and crops. A study of six silvo-arable site-system combinations across Europe found relative yield (wheat, barley, legumes) within

agroforestry systems in the range 50-80% compared to fields without trees. Impact on cereal and fodder crops increased with tree density (reduction of 20% in relative yield for every 100 trees/hectare), tree age (2.6% reduction in production per year tree age), and proximity to the trees (0.56% reduction in production with each additional metre towards the nearest tree). Although none of these models are precise, their results are corroborated by other studies, such as that of the mature, densely-tree-planted hardwood agroforestry system at Wakelyns by ORC in 2009, in which it was found that (at age 15 years), 50% decrease in wheat yields and 25% in oat yields compared to the centre of alleys. A similar study in 2012/13 of the short rotation coppice willow system showed that ley biomass at the edge of the alley was impacted. This effect was eliminated in the year following coppicing – potentially due to reduced nutrient and water competition as well as shade. Research by Cranfield University[19], meanwhile, estimated that 50% of grass production would be maintained over a 40-year period in a silvo-pastoral system of tree density 400 stems per hectare, and 74% of crop production would be maintained over 30 years in a silvo-arable system of 150 stems per hectare.

Such impacts may be reduced or overturned when the facilitatory effect of trees on crops arises, particularly in more stressed (hot, dry, windy) environments, as trees can positively influence microclimate (e.g., reducing evapotranspiration) and soil quality, nutrients and moisture[20]. Empirical evidence for such already exists in some British agroforestry systems, and indeed is a major motivation for agroforestry in a climate change context, with trees bringing environmental and economic resilience to such systems. Much more work is needed to model these adaptation benefits and understand their importance for the farm economy.

A combination of these facilitatory effects and principles of resource partitioning means that overall productivity (trees and crop production together) can be expected to be greater than the separated monoculture production of both, as agroforestry systems mature. This is commonly measured as Land Equivalent Ratio (LER) which typically falls within the range 1-1.4 (and sometimes up to 2.0[21]) when modelled or measured empirically. (Taking an average of 1.2, this means there is an overall 20% biomass yield advantage, meaning that you would need 20% more land in monocultures). An LER of 1.25 has been estimated at

19. The Potential Contribution of Agroforestry to Net Zero Objectives www.britishecologicalsociety.org/applied-ecology-resources/about-aer

20. Jose, S., 2009. *Agroforestry for ecosystem services and environmental benefits: an overview. Agroforestry systems*, 76, pp.1-10. And also: Ivezić, V., Yu, Y. and Werf, W.V.D., 2021. Crop yields in European agroforestry systems: a meta-analysis. Frontiers in Sustainable Food Systems, 5, p.606631.

21. Lehmann, L.M., Smith, J., Westaway, S., Pisanelli, A., Russo, G., Borek, R., Sandor, M., Gliga, A., Smith, L. and Ghaley, B.B., 2020. Productivity and economic evaluation of agroforestry systems for sustainable production of food and non-food products. Sustainability, 12(13), p.5429

Whitehall Farm, Peterborough, in this case consisting of a 24-metre-wide alley silvo-arable system with 86 apple trees/hectare.

As noted earlier, further potential benefits of agroforestry to the farm economy arise from a reduced need for inputs by, for example, provision of habitat for natural predators of crop pests, nutrient enrichment of soils, and tree fodder for livestock. While there is lower income from the reduced crop yields, the overall financial consequences of introducing agroforestry are dependent on the returns that can be achieved from the trees and shrubs: high value top fruit may be financially advantageous while low value biomass may result in reduced income. The value of both forage from grazed shrubs and shelter for livestock may be significant, but these have not been fully assessed and there are unquantified costs associated with the inconvenience of cropping between woodland strips.

Further reading

Wildlife and biodiversity: integration and management of farming and wildlife for their mutual benefit. IOTA Research Review: http://orgprints.org/5590/

Andrews, J. and Redbane, M. (1994) *Farming and Wildlife—A Practical Management Handbook.* Sold online still by most major distributors.

Chambers M, Crossland M, Westaway S, Smith J (2015) *A guide to harvesting woodfuel from hedges.* ORC Technical Guide. http://tinyurl.com/TWECOM-BPG

Lampkin, N. and Pearce, B. (2020) *Organic Farming and Biodiversity – Policy Options.* https://read.organicseurope.bio/publication/organic-farming-and-biodiversity/

Soil Association *Agroforestry Handbook* 2019 https://www.soilassociation.org/media/19141/the-agroforestry-handbook.pdf

Websites

Agricology: www.agricology.co.uk/resources/agroforestry/agroforestry-systems

Agroforestry for Livestock systems: SOLID Technical Note No 12
www.agricology.co.uk/sites/default/files/SOLID_technical_note12_agroforestry.pdf

Further information on tree planting, specifically focused on agroforestry:
www.soilassociation.org/causes-campaigns/agroforestry/

Defra guidance on agroforestry and support available for tree planting:
www.gov.uk/guidance/agroforestry-and-the-basic-payment-scheme

Woodland Trust

General information about tree planting with specific reference to grant schemes run by the Woodland Trust. www.woodlandtrust.org.uk/plant-trees/trees-for-landowners-and-farmers/

The role of trees in arable farming.
www.woodlandtrust.org.uk/publications/2015/04/role-of-trees-in-arable-farming/

How integrating trees can deliver large scale benefits on livestock farms.
www.woodlandtrust.org.uk/publications/2017/05/benefits-of-tree-shelter-for-livestock/

Benefits of trees on dairy farms – evaluating the use of trees on farms.
www.woodlandtrust.org.uk/publications/2013/11/benefits-of-trees-on-dairy-farms/

Planting trees to protect water - the role of trees and woods on farms. Woodland Trust.
www.woodlandtrust.org.uk/publications/2012/08/planting-trees-to-protect-water

SECTION 14: SOURCES OF FURTHER INFORMATION

The importance of information

Successful organic farming depends on accessing the best information available. Unlike conventional farming, which relies on inputs and the associated input advice, organic farming relies primarily on managing the farming system and using a farm's natural resource effectively. For farmers, having access to the experiences of other farmers and research derived information is very important. Therefore, there is a need for greater reliance on advisers, farmer groups, workshops, publications, and websites.

Providers of advice, information, training, and events

The Organic Research Centre (ORC) is the UK's leading independent organic research centre and is pioneering new approaches to plant breeding, cropping systems including agro-forestry, livestock production and environmental and economic issues, such as assessment of farm sustainability and of organic farm incomes and policies. ORC prioritises participatory research approaches where possible with the aim to build evidence and understanding of the positive impact of organic and agroecological farming, and practical information to help farmers and growers. Information on this work can be obtained through the ORC website (www.organicresearchcentre.com), the monthly E-Bulletin and through social media.

Agricology (www.agricology.co.uk) is a knowledge platform with a website at its core. It is focused on promoting and increasing the uptake of agroecological ways of farming and managing the land, and linking research with farmer practice, exploring practices that restore the farm ecosystem, such as reducing tillage to improve soil quality, planting 'cover crops', adding pollinator strips, and using agroforestry. The intention is to provide a toolbox for farmers and others in the industry to dip into and apply practically in various stages of their farming journey. Now funded and delivered by the ORC (originally founded in 2015 by the ORC, Daylesford Foundation and the GWCT), it encompasses the whole range of farming approaches; from conventional or integrated, to organic and biodynamic. It does this through collaboration with a wide range of different organisations from across the industry who contribute to the content that can be found on the website, collaborate on events, and generally help steer the direction of Agricology. The platform hosts farmer profiles or case studies, videos, podcasts, and a big library of resources on different practices and principles from across the sector. There are also guest blogs and research project pages, which provide a space for researchers and others involved in the industry to explore different topics and share their expertise.

Abacus Organic is a group of independent consultants, who work directly on farms and across the entire supply chain. Abacus works alongside CBs to engage farmers with the principles, regulation, and certification behind organic production. https://abacusagri.com/portfolio-items/organic-services/

CSA Network provide training, mentoring and support for Community Supported Agriculture (CSA) in the UK. Their resource library includes an A-Z guide, case studies and masterclasses. https://communitysupportedagriculture.org.uk/

Farm Carbon Toolkit work to further the understanding of greenhouse gas emissions in agriculture. They host a free Farm Carbon Calculator and Toolkit for farmers, and run an annual Soil Farmer of the Year competition, peer-to-peer training and learning events. They offer bespoke advice and action planning with farmers, landowners and companies, centred around measuring and improving soil health, greenhouse gas emissions and other ecosystem services. https://farmcarbontoolkit.org.uk/

Farm Health Online is a collaboration between Duchy College Rural Business School, Animal Welfare Approved and A Greener World. The website offers practical information to farmers for a range of diseases and species, on health management and animal welfare issues www.farmhealthonline.com/.

Garden Organic provides publications and information about organic growing, mainly aimed at gardeners, but the website contains information on weeds and their management, useful for farmers and growers. www.gardenorganic.org.uk.

Innovative Farmers is a not-for-profit membership network, for farmers and growers who are running on-farm trials to test innovative new practices. The website has a knowledge hub, a database searchable by topic and sector that contains results and guidance arising from field lab research.

Landworkers' Alliance (LWA) is a union of farmers, growers, foresters and land-based workers with a mission to improve the livelihoods of their members and create a better food and land-use system for everyone. They produce many useful publications regarding business planning and marketing, especially for new entrants. https://landworkersalliance.org.uk/

Organic Farmers and Growers is a certification body that also provides free technical information on its website (http://ofgorganic.org/). They also host the National Organic Conference 'an annual event built to service the stakeholder community in and around the organic sector'.

The Organic Growers Alliance (OGA) is a membership organisation supporting organic horticulture throughout the UK with a quarterly magazine, webinars, conferences and events (https://organicgrowersalliance.co.uk/). The members area of the website hosts back issues of The Organic Grower with an online index, an invaluable source of technical inormation.

Scottish Organic Producers Association Ltd. (SOPA) is Scotland's leading organic membership body, offering guidance and support to organic producers in Scotland, as well as access to certification: www.sopa.org.uk.

Scotland's Rural College (SRUC) engages in research and provides some advice information and for Scottish organic and converting producers. www.sruc.ac.uk/

Soil Association Farming and Land Use Team provides support and information to producers, such as the Organic Farming magazine, a national programme of training events and farm walks, initiatives in organic market development for UK producers. www.soilassociation.org

See Section 15 (Advice/Research) for full contact details.

Publications

Periodicals

Biological Agriculture and Horticulture. A scientific journal on every aspect of organic farming and sustainable production systems. Taylor & Francis.

Clover. Quarterly Magazine of the Organic Trust, Dublin.

Ecology and Farming. English language bulletin of international news and research reports on organic farming. IFOAM.

Growing Green International. Twice-yearly magazine published by the Vegan Organic Network (VON).

Living Earth. Triannual members' magazine on issues linking agriculture, food, health and the environment. Soil Association.

Organic Agriculture. The scientific journal of the International Society of Organic Agriculture Research ISOFAR. Springer.

Organic Farming. A technical quarterly magazine for producers. Soil Association.

Organic Matters. Bi-monthly newsletter on organic farming in Ireland. Irish Organic Association.

The Organic Grower. Quarterly magazine of the Organic Growers Alliance.

The Organic Way. A triannual membership magazine on organic gardening. Garden Organic.

Star and Furrow. Twice yearly journal for the Biodynamic Agricultural Association.

Books

Balfour E, (1946) The Living Soil Faber and Faber (Out of print. Free download available)

Blair, R (2016) A Practical Guide to the Feeding of Organic Farm Animals: Pigs, Poultry, Cattle, Sheep and Goats. 5m Publishing

Briggs S (2008) *Organic Cereal and Pulse Production*. Crowood Press, Marlborough.

Coleman E, (2018) *The New Organic Grower*, 30th Anniversary edition, Chelsea Green London

Cubison S (2009) *Organic Fruit and Viticulture*. Crowood Press, Marlborough.

Davies G, Lennartson M (2006) *Organic Vegetable Production – A Complete Guide*. Crowood Press, Marlborough. Out of print, but availableas e-book.

Davies G, Sumption P (2010) *Pest and Disease Management for Organic Farmers, Growers and Smallholders – A Complete Guide*, Crowood Press, Marlborough.

Davies G, Turner B (2008) Weed Management for Organic Farmers, Growers and Smallholders – A Complete Guide. Crowood Press, Marlborough.

Fiebrig I (Ed.) (2023) *Medicinal Agroecology*. CRC Press, Oxford.

Halberg N, Mueller A (2013): *Organic Agriculture for Sustainable Livelihoods*. Earthscan Food and Agriculture, Routledge.

Hall J, Tolhurst I (2007). *Growing Green: Animal-Free Organic Techniques*. Vegan Organic Network.

Howard A, (1940) An Agricultural Testament Oxford University Press (out of print but can be downloaded free on the internet)

Köpkie U., (2018) *Improving organic crop production*. Burleigh and Dodds.

Lampkin N, Padel S (1994) *Economics of Organic Farming*. CABI, Wallingford.

Lampkin N, (2002) *Organic Farming*. Old Pond, Ipswich (out of print)

Lockeretz W, (2007) *Organic Farming – an International History*. CABI, Wallingford.

Raskin B, Osborn S (Eds.) (2019) *The Agroforestry Handbook*. Soil Association Limited, Bristol.

Rundgreen G (2012) *Garden Earth*. Regeneration, Uppsala, Sweden. gardenearth.info/en/

Sumption P (2023) *The Organic Vegetable Grower – A Practical Guide to Growing for the Market*, Crowood Press, Marlborough. (In press)

Tolhurst I:(2016) *Back to Earth: UK Organic Horticulture through the Lifetime of a Grower*. Tolhurst Organic Partnership.

Vaarst M et al (Eds.) (2004) *Animal Health and Welfare in Organic Agriculture*, CABI, Wallingford.

Vaarst M, Roderick S (Eds.) (2019) *Improving organic animal farming*. Burleigh Dodds Science Publishing, Cambridge.

Younie D (2012) *Grassland Management for Organic Farmers*. Crowood Press, Marlborough.

Various Technical Guides and Research Reviews published by IOTA, Garden Organic, Organic Centre Wales, Organic Farmers and Growers, Organic Research Centre and Soil Association (see footnotes in various sections for titles/links and Section 15 for contact details). Many are accessible via https://agricology.co.uk/

Websites

AFINET and AGFORWARD agroforestry factsheets: https://agroforestrynet.eu/afinet/handbook

Agricology https://agricology.co.uk/ See specifically:

- Organic Management Techniques Project outputs: https://agricology.co.uk/research-projects/organic-management-techniques-project/
- Farm System Health in Practice outputs: https://agricology.co.uk/research-projects/farm-system-health-practice/

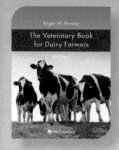

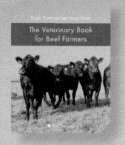

DARDNI Organic Production: www.daera-ni.gov.uk/topics/livestock-farming/organic-farming

Defra Organic Farming: www.gov.uk/organic-systems-and-standards-in-farming

- www.gov.uk/guidance/organic-farming-how-to-get-certification-and-apply-for-funding

EU Organic Farming Pages: http://ec.europa.eu/agriculture/organic/index_en

FAO Organic Agriculture: www.fao.org/organicag

FiBL Organic Research: www.fibl.org/en/

Garden Organic: www.gardenorganic.org.uk/

IFOAM International Federation of Organic Agriculture Movements: www.ifoam.bio

IFOAM Organics Europe: www.organicseurope.bio/

Organic Europe: www.organic-europe.net

Organic Farm Knowledge (including OK-NET Arable and OK-NET EcoFeed): https://organic-farmknowledge.org/

Organic Research Centre: www.organicresearchcentre.com

Organic X Seeds: www.organicxseeds.com

Soil Association: www.soilassociation.org and www.soilassociation.org/our-work-in-scotland/

Teagasc (Ireland): www.teagasc.ie/rural-economy/organics/

USDA Alternative Farming Systems Information Centre: www.nal.usda.gov/afsic

Whole Health Agriculture (WHAg): https://wholehealthag.org/

SECTION 15: ADDRESSES

Organic certification bodies – see Section 3 for details

Organic membership organisations

Biodynamic Association (BDA)

Open House, Gloucester Street, Stroud, GL5 1QG
Tel: 01453 759501 (General) 01453 766296 (Certification)
Internet: www.biodynamic.org.uk Email: office@biodynamic.org.uk

Farm Retail Association

c/o Yorkshire Agricultural Society, Great Yorkshire Showground, Harrogate, HG2 8NZ
Tel. 01423 546214
Internet: https://farmretail.co.uk/ Email: laura@farmretail.co.uk

Garden Organic

Ryton Gardens, Wolston Lane, Coventry, Warwickshire, CV8 3LG
Tel: 024 7630 3517
Internet: www.gardenorganic.org.uk Email: enquiry@gardenorganic.org.uk

IFOAM – Organics International (IFOAM)

Head Office, Charles-de-Gaulle-Street 5, 53113 Bonn, Germany
Tel: +49 (0) 228 926 50-10
Internet: www.ifoam.bio Email: contact@ifoam.bio

Irish Organic Association (IOA)

Unit 13 Inish Carrig, Golden Island, Athlone, Co. Westmeath, Ireland N37 N1W4
Tel: +353 (0) 90 643 3680
Internet: https://www.irishorganicassociation.ie/ Email: info@irishoa.ie

Organic Growers Alliance (OGA)

187 Winsley Road, Bradford on Avon, Wiltshire, BA15 1NX
Internet: www.organicgrowersalliance.co.uk Email: hello@organicgrowersalliance.co.uk

Organic Trade Board (OTB)

18 Rooksbury Avenue, Andover, SP10 2LW
Internet: www.organictradeboard.co.uk Email: hello@organictradeboard.co.uk

Scottish Organic Producers Association (SOPA)

c/o Drummond Laurie, Algo Business Centre, Glenearn Rd, Perth, PH2 8BX. Tel: 0300 772 9795
Internet: www.sopa.org.uk Email: info@sopa.org.uk

Soil Association

Spear House, 51 Victoria Street, Bristol, BS1 6AD
Tel: 0300 330 0100
Internet: www.soilassociation.org

Soil Association Scotland

20 Potterrow, Edinburgh, EH8 9BL
Tel: 0131 370 8150
Internet: www.soilassociation.org/scotland

World Wide Opportunities on Organic Farms WWOOF UK

61 Bridge Street, Kington, Herefordshire, HR5 3DJ
Internet: www.wwoof.org.uk Email: info@wwoof.org.uk

Organic advice and research

See also Section 14 for a short description of some of the most important organisations.

Abacus Organic

Ian Knight, Tel: 07775 842444 Stephen Briggs, Tel: 07855 341309
Internet: https://abacusagri.com/portfolio-items/organic-services/
Email: advice@abacusagri.com

Agricology

Trent Lodge, Stroud Road, Cirencester, Gloucestershire, GL7 6JN
Internet: www.agricology.co.uk Email: enquiries@agricology.co.uk

College of Agriculture, Food and Rural Enterprise (CAFRE)

Greenmount Campus, 45 Tirgracy Road, Antrim, Co. Antrim, BT41 4PS, Northern Ireland
Tel: 028 9442 6601
Internet: www.cafre.ac.uk

Farm Health Online

Internet: www.farmhealthonline.com Email: farmhealth@duchy.ac.uk

Farming and Wildlife Advisory Group FWAG

See contact section on website for regional contacts.
Internet: www.fwag.org.uk Email: hello@fwag.org.uk

Farming Connect Wales

Operated by Business Wales.
Tel: 08456 000 813
Internet: www.businesswales.gov.wales/farmingconnect
Email: use online contact form or see website for regional contacts

Game and Wildlife Conservation Trust (GWCT)

Allerton Project, Loddington House, Main Street, Loddington, Leicestershire, LE7 9XE
Tel: 01572 717220
Internet: www.gwct.org.uk/allerton/ Email: allerton@gwct.org.uk

Institute of Biological, Environmental and Rural Sciences (IBERS)

Aberystwyth University; Penglais Campus, Aberystwyth, Ceredigion, SY23 3EB
Tel: 01970 823000
Internet: www.aber.ac.uk/en/ibers Email: ibrstaff@aber.ac.uk

Nafferton Ecological Farming Group (Newcastle University)

Nafferton Farm, Stocksfield, Northumberland, NE43 7XD
Tel: 01661 830222
Internet: www.nefg-organic.org Email: contact@nefg-organic.org

National Institute of Agricultural Botany NIAB

93 Lawrence Weaver Road, Cambridge, Cambridgeshire, CB3 0LE
Tel: 01223 342200
Internet: www.niab.com Email: info@niab.com

Organic Centre (Ireland)

Rossinver, Co. Leitrim, Ireland
Tel: +353 071 985 4338
Internet: theorganiccentre.ie Email: info@theorganiccentre.ie

Organic Research Centre

Trent Lodge, Stroud Road, Cirencester, Gloucestershire GL7 6JN
Tel: 01488 658298
Internet: www.organicresearchcentre.com Email: complete online contact form

Pesticides Action Network (PAN UK)

The Brighthelm Centre, North Road, Brighton, BN1 1YD
Tel: 01273 964 230
Internet: www.pan-uk.org Email: admin@pan-uk.org

RSK ADAS-UK

Spring Lodge, 172 Chester Road, Helsby, Cheshire, WA6 0AR
Tel: 0333 0142950
Internet: https://adas.co.uk/

SAC Consulting (SRUC)

Internet: www.sruc.ac.uk/business-services/sac-consulting/contact-sac-consulting/
Contact: visit website to see regional consulting offices in Scotland and Northern England

Scotland's Rural College (SRUC)

Peter Wilson Building, The King's Buildings, West Mains Road, Edinburgh, EH9 3JG
Internet: https://www.sruc.ac.uk/ Email: complete online contact form

Sustain

The Green House, 244-254 Cambridge Heath Road, London E2 9DATel: 020 3559 6777
Internet: www.sustainweb.org Email: sustain@sustainweb.org

Teagasc (Ireland)

Oak Park, Carlow, Co. Carlow, Ireland
Tel: +353 (0) 59 917 0200
Internet: www.teagasc.ie Email: info@teagasc.ie

Warwick Crop Centre

School of Life Sciences, University of Warwick, Innovation Campus, Stratford-upon-Avon, CV35 9EF
Internet: https://warwick.ac.uk/fac/sci/lifesci/wcc/
Email: cropcentre@warwick.ac.uk

Organic education and training

The following listings include only places that offer specialist courses or short courses in organic agriculture or closely related subjects.

Abacus Organic

Ian Knight, Tel: 07775 842444 Stephen Briggs, Tel: 07855 341309.
Internet: www.abacusagri.com/organic-farming-services/
Email: advice@abacusagri.com

- Seminars, courses, workshops and discussion groups for farmers

ACS Distance Education

P.O. Box 4171, Stourbridge, DY8 2WZ.
Tel: 01384 442752
Internet: www.acsedu.co.uk Email: info@acsedu.co.uk

- Certificate in Horticulture (Organic plant growing) VHT002
- Certificate in Commercial Organic Vegetable Growing VHT241
- Organic Farming BAG305 (correspondence)
- Organic Plant Culture BHT302 (correspondence)
- Advanced Permaculture BHT301 & Permaculture Consulting VHT37(correspondence)

Bangor University

Bangor, Gwynedd, LL57 2DG, UK

Tel: 01248 351151

Internet: www.bangor.ac.uk/ Email: admissions@bangor.ac.uk

- MSc Agroforestry and food security

B.E.S.T in Horticulture Education (RHS Approved Centres in the Midlands and Oxfordshire)
Internet: www.bestinhorticulture.co.uk/

- Certificate in Organic Horticulture (correspondence)

Biodynamic Agricultural College (formerly Emerson College)

The Old Painswick Inn, Gloucester Street, Stroud, GL5 1QG.
Tel: 01453 766296
Internet: www.bdacollege.org.uk Email: info@bdacollege.org.uk

- Biodynamic Principles and Practice
- Diploma in Biodynamic Farming and Growing (work based)

Centre for Agroecology, Water and Resilience, Coventry University

Priory Street, Coventry CV1 5FB
Internet: www.coventry.ac.uk/research/areas-of-research/agroecology-water-resilience/

- Agroecology, Water and Food Sovereignty MSc

Centre for Alternative Technology

Machynlleth, Powys, SY20 9AZ, Tel: 01654 705950 (General enquiries)
Internet: www.cat.org.uk Email: complete online form

- Specialist short courses

College of Agriculture, Food and Rural Enterprise (CAFRE)

Greenmount Campus, 45 Tirgracy Road, Antrim, BT41 4PS. Tel: 0 028 9442 6601
Internet: www.cafre.ac.uk   Email: complete online form
- Principles of Organic Horticulture (Distance learning)
- Introductory courses in organic production – (2 day course)

Duchy College

Internet: www.duchy.ac.uk   Email: complete online form
- Range of courses, from apprenticeships to short and adult courses – visit website to find colleges in Cornwall and Devon

Homoeopathy at Wellie Level (HAWL)

Tel: 01666 841213
Internet: www.hawl.co.uk   Email: secretary@hawl.co.uk
- Homeopathy at Wellie Level (3 days, open to all). Various locations – in-person course delivered in Tetbury GL8 8NF but online course available.

Horticulture Correspondence College

Internet: www.hccollege.co.uk/index.php/course/organic-gadening
- Organic Gardening (QLS Level 3)

National Organic Training Skillnet (NOTS Ireland)

The Enterprise Centre, Hill Road, Drumshanbo, Co. Leitrim, Ireland
Tel: 071 96 40688
Internet: www.nots.ie   Email: info@nots.ie
- Various short courses for farmers and growers

University of Newcastle Upon Tyne

Newcastle Upon Tyne NE1 7RU
Tel: 0191 222 5594/27
Internet: www.ncl.ac.uk   Email: Use online contact form
- MSc Sustainable Agriculture and Food Security (f/t)

Organic Centre (Ireland)

Rossinver, Co. Leitrim, Ireland
Tel: +353 071 985 4338 Internet: theorganiccentre.ie   Email: info@theorganiccentre.ie
- Full Time Organic Horticulture Training Course
- Range of short courses

The Organic College,

Dromcollogher, Co. Limerick, Ireland.
Tel: +353 (0)63 83604
Internet: www.organiccollege.com   Email: oifig@organiccollege.com
- Certificate Organic Growing and Sustainable Living Skills (with options including agriculture, horticulture and sustainable development) (f/t 1 year and p/t options)
- Short courses including workshops in organic food and farming on topics such as soil science, organic standards and plant protection

Organic Research Centre
Trent Lodge, Stroud Road, Cirencester, Gloucestershire, GL7 6JN. Tel: 01488 658298
Internet: www.organicresearchcentre.com   Email: complete online form
- Various short courses for farmers and growers"

University of Reading

Whiteknights House, Reading RG6 6UR
Tel: +44 (0) 118 378 5289 (PG admissions).
Internet: www.reading.ac.uk/ Email: pgadmissions@reading.ac.uk

- Post graduate courses include MSc of Research Agriculture, Ecology and Environment, amongst other relevant undergraduate and postgraduate courses

Royal Agricultural University

Cirencester, Gloucestershire, GL7 6JS. Tel: 01285 652531
Internet: www.rau.ac.uk

- Modules on integrated agroecological systems,production and marketing

Schumacher College

The Old Postern, Dartington, Totnes, Devon TQ9 6EA.
Tel: 01803 865934
Internet: https://campus.dartington.org/schumacher-college/
Email: admissions@dartington.org

- Various short, vocational and postgraduate courses

Scotland's Rural College (SRUC)

Peter Wilson Building, The King's Buildings, West Mains Road, Edinburgh, EH9 3JGTel: 0131 535 4000 (general enquiries) Internet: www.sruc.ac.uk
Email: complete online form

- Various short and full time courses

Teagasc (Ireland)

Oak Park, Carlow, Co. Carlow, Ireland.
Tel: +353 599170200
Internet: www.teagasc.ie Email: info@teagasc.ie

- Variety of agriculture and land management short courses
- Organic Farming Principles (Level 5 QQI 25 hours)

Warwickshire College Group

Pershore Campus, Avonbank, Pershore, WR10 3JP.
Tel: 0300 4560047
Internet: www.warwickshire.ac.uk Email: info@wcg.ac.uk

- Distance learning courses in organic farming and related
- BSc Sustainable Horticultural Technology

European courses:

EUR Organic

Internet:www.eur-organic.eu/en

- EUR-Organic is a unique study programme in Europe offered by five renowned universities. It combines high-ranking research environments with state-of-the-art knowledge from industry leaders in organic food production.

Food and Agriculture Organisation (FAO) Training Manual for Organic Agriculture

https://tinyurl.com/FAO-organic-training

FiBL

www.fibl.org/en/themes/education-training

Governmental grant-awarding agencies and advisory services

Business Wales

Many local centres see website for local details. Tel: 03000 603 000 (main)
Internet: www.businesswales.gov.wales

Business Gateway - Scotland

Many local centres. See website and separate entry for Highlands and Islands.
Tel: 0300 013 4753 (main) Internet: www.bgateway.com

Crofting Commission

Great Glen House, Leachkin Road, Inverness, IV3 8NW. Tel: 01463 663439
Internet: www.crofting.scotland.gov.uk Email: complete online form

Department for Environment, Food and Rural Affairs (Defra)
Seacole Building, 2 Marsham Street, London, SW1P 4DF
Tel: 03000 200 301 (Rural services helpline)
Internet: www.gov.uk/defra Email: defra.helpline@defra.gsi.gov.uk

Department of Agriculture, Fisheries and Food (Republic of Ireland)

Head Office, Agriculture House, Kildare Street, Dublin 2. Tel: +353 (0) 1607 2000
Internet: www.agriculture.gov.ie Email: info@agriculture.gov.ie

Department of Agriculture, Environment and Rural Affairs (DAERA)

Dundonald House, Upper Newtownards Road, Belfast, BT4 3SB. Tel: 0300 2007852
Internet: www.daera-ni.gov.uk Email: daera.helpline@daera-ni.gov.uk

Environment Agency

PO Box 544, Rotherham, S60 1BY
Tel: 03708 506 506
Internet: www.environment-agency.gov.uk Email: enquiries@environment-agency.gov.uk

Farming Connect Wales Service Centre

Welsh Government, Rhodfa Padarn, Llanbadarn Fawr, Aberystwyth, SY23 3UE.
Tel: 08456 000813 (Main), Regional contacts listed on website
Internet: https://businesswales.gov.wales/farmingconnect/ Email: complete online form

Forestry Division - Republic of Ireland

Department of Agriculture, Food and the Marine (DAFM), Head Office, Agriculture House, Kildare Street, Dublin 2. Tel: +35 (0)53 91 63400. Lo-call: 0761 064 415
Email: forestryinfo@agriculture.gov.ie

Forestry England

620 Bristol Business Park, Coldharbour Lane, Bristol, BS16 1EJ.
Tel: 0300 067 4000 Internet: https://www.forestryengland.uk/
Email: enquiries@forestryengland.uk

Forest Service (Northern Ireland)

Inishkeen House, Killyhevlin, Enniskillen, BT74 4EJ. Tel: +02866 343 165
Internet: https://www.daera-ni.gov.uk/forest-service
Email: customer.forestservice@daera-ni.gov.uk

Highlands and Island Enterprise

An Lòchran, 10 Inverness Campus, IV2 5NA. Tel: 01463 245 245
Internet: www.hie.co.uk Email: enquiries@hient.co.uk

Invest Northern Ireland

Tel: 0800 181 4422. Regional office details on 'Contact us' page on website.
Internet: www.investni.com Email: Use online contact form

Natural England

County Hall, Spetchley Road, Worcester, WR5 2NP
Tel: 0300 060 3900
Internet: www.gov.uk/government/organisations/natural-england
Email: enquiries@naturalengland.org.uk
For Grants: www.gov.uk/topic/farming-food-grants-payments/rural-grants-payments

Natural Resources Wales

Rivers House, St. Mellons Business Park, Fortran Rd, Cardiff CF3 0EYTel: 0300 065 3000
Internet: www.naturalresources.wales Email: enquiries@naturalresourceswales.gov.uk

Rural Payments Agency (RPA)

PO Box 69, Reading, RG1 3YD.
Tel: 03000 200 301 (Main)
Helpline: Cattle Tracing System (CTS) 0345 050 1234 (general) and 0345 050 3456 (Welsh)
Internet: www.gov.uk/government/organisations/rural-payments-agency
Email: ruralpayments@defra.gsi.gov.uk

Rural Payments and Services (Scotland)

Tel: 0300 300 2222 (Entitlements and Payments information), Local office details online.
Internet: www.ruralpayments.org/publicsite/futures/topics/

Rural Payments Wales

RPW, PO Box 1081, Cardiff, CF11 1SU. Tel: 0300 062 5004
Internet: gov.wales/topics/environmentcountryside/farmingandcountryside/rpwonline.

Scottish Enterprise

Atrium Court, 50 Waterloo Street, Glasgow, G2 6HQ5.
Tel: 0300 013 3385
Internet: www.scottish-enterprise.com Email: Use contact form on website

Scottish Forestry

Saughton House, Broomhouse Drive, Edinburgh, EH11 3XD. Tel: 0131 370 5250
Internet: https://forestry.gov.scot/ Email: scottish.forestry@forestry.gov.scot

Scottish Government (Agriculture and Rural Economy)

Internet: www.gov.scot/Topics/farmingrural
Directory:www.gov.scot/about/how-government-is-run/directorates/agriculture-rural-economy/

Scottish Natural Heritage (SNH)

HQ: Great Glen House, Leachkin Road, Inverness, IV3 8NW
For Regional addresses & telephone numbers see website
Tel: 01463 725000
Internet: www.snh.gov.uk Email: enquiries@snh.gov.uk

Welsh Government Farm Liaison Service
See https://www.gov.wales/contact-farm-liaison-service for regional office details.
Email: farmliaisonservice@gov.wales.

Welsh Government – Food and Drink Wales

Internet: https://businesswales.gov.wales/foodanddrink/ Tel: 03000 6 03000

SECTION 15
ADDRESSES

A

B

C

SECTION 16 INDEX

H

I

L

M

N

O

P

Q

R

S

T

U

V

W

Y